Confederate Torpedoes

Confederate Torpedoes

Two Illustrated 19th Century Works with New Appendices and Photographs

Gabriel J. Rains *and*
Peter S. Michie

EDITED BY HERBERT M. SCHILLER

McFarland & Company, Inc., Publishers
Jefferson, North Carolina, and London

Gabriel J. Rains (1803–1881)
Peter S. Michie (1838–1901)

LIBRARY OF CONGRESS CATALOGUING-IN-PUBLICATION DATA

Rains, Gabriel James, 1803–1881.
Confederate torpedoes : two illustrated 19th century works
with new appendices and photographs / Gabriel J. Rains
and Peter S. Michie ; edited by Herbert M. Schiller.
p. cm.
Includes bibliographical references and index.

ISBN 978-0-7864-6332-9
softcover : 50# alkaline paper ♾

1. Submarine mines — Confederate States of America.
2. Mines (Military explosives) — Confederate States of America.
3. United States — History — Civil War, 1861–1865 — Naval
operations, Confederate. 4. United States — History — Civil War,
1861–1865 — Equipment and supplies. 5. Rains, Gabriel James,
1803–1881. 6. Michie, Peter Smith, 1839–1901. I. Michie, Peter
Smith, 1839–1901. II. Schiller, Herbert M., 1943– III. Title.
V856.5.U6R35 2011 623.4'5115 — dc22 2011002723

BRITISH LIBRARY CATALOGUING DATA ARE AVAILABLE

Front cover: "Magneto Electric Torpedo," plate 11 from
Gabriel J. Rains' *Torpedo Book* (see page 52 for a full caption).

Manufactured in the United States of America

*McFarland & Company, Inc., Publishers
Box 611, Jefferson, North Carolina 28640
www.mcfarlandpub.com*

In honor of my father, Jacob "Jack" Schiller,
a mechanical engineer who early in his career
was involved in the manufacture of torpedoes.
—*The Editor*

Contents

Acknowledgments

I first learned of the details of the Rains manuscript in 1989 when I read William Davis Waters's article in the *North Carolina Historical Review*. I had read Milton F. Perry's *Infernal Machines* several years before, but hadn't paid much attention to the references. I tore out Water's article and stuck it in my copy of Perry's book, and there it remained for twenty years. Although I had retired in 2007, I never thought of writing another book. For some reason, I pulled out the article, read it again, and contacted John Coski at The Museum of the Confederacy about the possibility of transcribing the text and producing digital images of the plates, with the sole initial intention of having the fun of working with the manuscript.

During a visit with John, he mentioned Mike Kochan's little book on torpedoes, which I purchased and found to be very informative. Among his references was Peter S. Michie's manuscript and plates, at the United States Military Academy library.

Upon calling the switchboard of the Academy, I was given Suzanne Christoff's telephone number, and when I called I left a message about what I was working on, i.e., Rains's *Torpedo Book*, and that I had some questions about his proposal to teach a course on torpedoes there, and by the way, what about the Michie manuscript? Suzanne didn't know anything about the Rains book, and thought I must be confusing it with the Michie material. One thing led to another and this project was born.

I thought the juxtaposition of the Rains's manuscript, dealing primarily with the construction and deployment of torpedoes by the Confederate forces with Michie's report, written shortly after the end of the Civil War, about his findings in Richmond at the Torpedo Bureau laboratories, would not only be an interesting combination but also would make primary source material on this subject available to a broader audience than it had been before.

First and foremost I acknowledge the help and encouragement of John Coski of The Museum of the Confederacy. He made the original Rains manuscript available to me, and along with the staff of the manuscript room, assisted and encouraged me in every way. Second is Suzanne Christoff, Chief Librarian of the library of the United States Military Academy at West Point. Although she, along with Valarie Dutdut and Susan Lintelmann, was involved in a relocation of the entire USMA library at the time, they all took an interest in my project and had copies of material made available to me.

Les Jensen provided me with copies of the accession cards for the various torpedoes in the museum collection at the United States Military Academy. He further provided me with scanned photographs of the various specimens.

In September, 2009, I visited the Military Academy. Suzanne, Valarie, and Susan had prepared everything for me, including Michie's manuscript and plates, a copy of W.R. King's *Torpedoes: Their Invention and Use* (discussed in the introduction to the Michie manuscript), and manuscript files on Rains and Michie. Valarie had already located the letter from the War Department to Rains declining his offer to teach a course on torpedoes at the Military Academy (discussed in detail in the introduction to the Rains manuscript). Les Jensen made time available to photograph the torpedoes. I arrived concerned that the glass cases would cause problems with reflections and concerned that I couldn't do a better job than the photographs he had supplied for me. What a surprise when he lead me into the sub-basement vaults, through two separate steel doors, into a vast Aladdin's cave of military artifacts. Over two hours he and I moved and dusted the torpedoes, one by one, to a photography area, and I could take all the images I wanted. Pure exhilaration!

David Lowe, foremost proponent of *humo veritas*, kindly took his time to read the draft of this manuscript and offered many helpful suggestions. The remaining errors are my sole responsibility.

Gordon Jones, Senior Military Historian and Curator of the Atlanta History Center, kindly provided me with multiple images of the subterra shell in their collection, as well as an original sensitive fuse with safety cap. Stephen Crowell, Curator of the Naval Undersea Museum, provided me with multiple images of their Confederate frame torpedo. Jimmy Blankenship, historian and curator of the Hopewell Unit of the Petersburg National Battlefield, provided me with photographs and information about the horological torpedo detonator. Ray Flowers, historian at Fort Fisher, kindly provided me with photographs of torpedo specimens on display and helped to answer innumerable questions. Lonnie Looper, whom I had never met, spent a weekend afternoon working with me in providing images in various formats of his replica subterra shell. It never ceases to amaze me how helpful people will be to others with similar interests. Mike Kochen shared information, images, and contacts. His wonderful book, cited in the bibliography, will be helpful to anyone with an interest is the more technical aspects of the subject of *torpedoes*. His associate, John Wideman, kindly shared their research on subterra shells and horological mechanisms.

Everyone who does any work on the subject of Confederate torpedoes owes a tremendous debt to Milton F. Perry, whose 1965 *Infernal Machines* is indispensable. Anything that I have accomplished has been done standing on his shoulders.

DeAnne Blanton, Nik Butler, Pat Flynn, Chris Fonvielle, Diane Jacob, Jan Hensley, Earl Hess, Bob Krick, Grahame Long, Jim McKee, Randy Watkins, and Jeanne West all helped answer various questions which arose about events, places, and names. I am grateful for all the help I received with this work, and especially for my patient wife, Annette, who knows all too well how I "am" when involved in yet another "project."—H.M.S.

Preface

I had several goals in preparing this material. The first was to prepare a usable, accurate transcript of the remarkable Gabriel Rains *Torpedo Book*, with its amazing plates, few of which have ever been published. Although it is not the official document used in the wartime manufacture of Confederate torpedoes, it is the next best thing. The second goal was to present Peter Michie's report, portions of which have never been published. He proved an able observer on the scene immediately after the fall of Richmond in 1865 when he visited the Torpedo Laboratory to collect specimens and talk to a former worker, who explained the mechanisms of the these mysterious instruments. With staff and time available, Michie produced twenty-one accurate hand-colored plates, each over two feet square, of these deadly instruments. Although the plates were rendered as small engravings in 1866, the originals have never been reproduced before.

My third goal was to provide new photographs of the torpedo specimens collected at the United States Military Academy Museum, as well as to locate specimens in other museums and in private collections, and provide clear images of examples of some of the many devices created by the Confederates during the Civil War. Finally, I have tried to expand and correct the lists of vessels damaged and sunk by Confederate torpedoes, by reviewing the source documents to provide context to the events surrounding each episode.

I have tried not to stray into areas where I know others are doing research, i.e., the history of the use of subterra shells in the Civil War, and a recounting of the evolution of the technical and manufacturing aspects of torpedo construction. I hope that the material presented here will help and encourage others to recount more details about the manufacture and deployment of these devices, which certainly inspired respect, if not fear, among the members of the wartime United States Navy.—H.M.S.

I. Gabriel J. Rains

Biographical Introduction

Gabriel James Rains is primarily remembered for his work with torpedoes; I propose here to summarize that work along with a brief account of both the Confederate Army and Navy endeavors with water and land mines (then called torpedoes).[1] Rains was born in New Bern, North Carolina, in 1803, one of eight children. His father, Gabriel Maginault Rains, a French Huguenot, was successful as a cabinet maker and upholsterer. Young Rains attended the local academy, graduating at the age of seventeen. Three years later, in 1823, he began his course of instruction at the United States Military Academy at West Point, New York.

His class began with ninety-eight cadets; when he graduated four years later he was ranked thirteenth among the remaining thirty-eight cadets. He excelled in chemistry, artillery, and drawing. Jefferson Davis, Robert E. Lee, and Joseph E. Johnston were his friends among the upperclassmen.

As a second lieutenant in 1827 he began twelve years' service in the Indian Territory. For part of that time he worked with Lieutenant Jefferson Davis and their friendship apparently grew. In 1835, during his frontier service, he married Mary Jane McClellan, granddaughter of John Sevier, who had served as a North Carolina congressman and governor of Tennessee. In 1839, Captain Rains was transferred to Fort King in north central Florida, where conflict had erupted between the white settlers and the Seminole Indians. He commanded Company A of the 7th Infantry.[2] In 1840, during the Second Seminole War (1835–1842), Rains contrived a booby-trap using an explosive shell, probably with a friction primer, buried in the soil, to alert his force of approaching Indians. A subsequent explosion frightened the Indians away, but when Rains and a group of men went to investigate the explosion, he was seriously wounded in the chest by other Seminoles. Although the wound was initially thought to be mortal, Rains recovered and was promoted to brevet major for his conduct under fire.

Five years later, Rains saw combat in Mexico and was then assigned by General Zachary Taylor to recruiting duty. Following the conclusion of hostilities in Mexico, he returned to Florida and was promoted to major. After another recruiting assignment in New York, he was sent in June, 1853, to more frontier duty in the territory of Washington, and three years later as garrison commander of Fort Humboldt in the California Bay, with the rank of lieutenant colonel. Finally in 1860 Lieutenant Colonel Rains was assigned to duty in Vermont, where he learned during the following year that his native North Carolina had seceded. On July 31, 1861, he resigned his commission in the United States Army and tendered his service to the new Confederacy.[3]

That autumn Confederate president Jefferson Davis personally recommended that Rains be appointed a Brigadier General.[4] In October, 1861, he arrived in Yorktown, Virginia, and took command of a brigade of the Army of the Department of the Peninsula, where he busied himself with garrison duties and defense preparation. In early April, 1862, Major

General George B. McClellan began his ponderous movement of Union forces up the Virginia peninsula towards Richmond.

During the month the Federal commander labored to prepare works for a siege at Yorktown, and when McClellan was almost ready on May 3 to begin operations, the Confederate forces, including Rains', began their withdrawal. To delay the advancing Federal forces, Rains' troops planted a number of "torpedoes," or subterra shells, along their evacuation route to Williamsburg. After a brief engagement with pursuing Federal troops outside Williamsburg, Rains resumed his withdrawal toward Richmond, again leaving torpedoes behind him to slow his pursuers. By now unsuspecting Union cavalrymen, moving through the abandoned Confederate works in Yorktown, exploded some of the subterranean shells, killing several men. General McClellan condemned the torpedoes as "murderous and barbarous" and threatened to use Confederate prisoners to remove them.[5] The Confederate commanding general, Joseph E. Johnston, launched an investigation into the use of the torpedoes; Gains readily explained his use of the devices to slow the advancing enemy.[6] Although Johnston condemned the practice, Rains persisted in justifying the advantages of using the subterranean shells in both deterring and demoralizing the enemy.[7] The matter was subsequently reviewed by the Confederate Secretary of War, George W. Randolph, who did not support the use of the subterra shells "merely to destroy life" by killing a few men without materially injuring the enemy's force, but did support the use of torpedoes to save defensive works or to defend armies from attackers or pursuers. To resolve the disagreement between Rains and his superior officers, Rains was given the option to transferring to work on river defenses, where the use of such devices to defend a river or harbor to deter attacking or blockading fleets was "admissible."[8] Rains, however, remained with the army and his brigade took part in the battle at Seven Pines on May 31, 1862, although his performance was judged as somewhat lacking by his immediate superior, Major General D.H. Hill.[9] At that same battle General Johnston was seriously wounded, and General Robert E. Lee, who had previously served at President Davis's military advisor, was appointed commander of the Confederate forces, now east of Richmond. As part of Lee's evaluation of his defensive situation, he summoned Rains to discuss the latter's new role. Lee was concerned that the Federals would use the James River to bring more troops toward Richmond, and on June 18, 1862, Rains was assigned to protect the James and Appomattox Rivers using "submarine defenses."[10]

Gabriel J. Rains (United States Military Academy)

The concept of using defensive mines in the rivers of the Confederacy was not new. During mid–June, 1861, Commander Matthew F. Maury, who had achieved prominence in nautical affairs for the United States Navy, including describing the Gulf Stream earlier in his career, had demonstrated a marine explosive device at the Richmond Wharf for Secretary of the Navy Stephen Mallory.[11] The Naval Bureau of Coast, Harbor, and River Defense was authorized and Maury was named director. The Bureau developed mines that were used in the Mississippi River and near Columbus, Kentucky,[12] and in 1862 Maury's bureau began work on obstructing the James River. On June 19, he reported that 15 electric mines, four containing 160 pounds of gunpowder, the remainder containing 70 pounds, were in the river in the vicinity of Chaffin's Bluff, downstream from the guns of Fort Darling on Drewry's Bluff.[13] In mid–June, 1862, Maury was reassigned to Europe to oversee construction of vessels for the Confederate Navy as well as to procure naval supplies, to include wire and material for torpedoes. Lieutenant Hunter Davidson, Maury's chief aide, was appointed to replace him on June 20, 1862.[14]

For the next three months Rains developed and placed "submarine mortar batteries" in the James River. These contrivances used readily available material — artillery shells, timber, and contact primers — and were designed to be detonated by contact with the bottom of a ship's hull. They were unlike those being developed by the navy which required scarce insulated, waterproof wire and galvanic cells to generate electric current and an operator to discharge the weapon when the enemy vessels seemed to be in the correct position. Rains also developed the simple barrel torpedo, which also only required contact with a vessel's hull to explode.[15]

Upon the termination of McClellan's peninsula campaign in August 1862, Rains was ordered to report to Wilmington, North Carolina, and the Navy took over mining operations on the James River. In the District of the Cape Fear, Rains served primarily in an administrative capacity while planning a system of torpedo deployment in the lower Cape Fear River.[16] He would remain at Wilmington until early December 1862, when he returned to Richmond.

Meanwhile, the Confederate Congress, on October 25, 1862, had authorized two new offices, the Naval Submarine Battery Service, headed by Lieutenant Hunter Davidson, and the Army Torpedo Bureau, headed by General Rains.[17] The existence of the Torpedo Bureau was intended to be kept secret, and when Rains returned to Richmond he ostensibly headed the Bureau of Conscription, work he was familiar with from his time in the old army, while in secret he continued his work with torpedoes.[18] Lieutenant Colonel George W. Lay handled the routine daily activities of the Conscription Bureau.[19] During this time Rains continued to work on his sensitive primer fuse. He initially wanted the primer to detonate from the "slightest pressure," but after inflicting irreparable damage to his forefinger and thumb of his right hand on January 24, 1863, he settled on a detonation pressure of seven pounds.[20]

In May 1863 Rains was relieved of his duties at the Conscription Department so he could spend all his time developing his various devices. He would eventually have torpedo production facilities not only in Richmond and Charleston, but also Wilmington, North Carolina, Savannah, Georgia, and Mobile, Alabama. It was around this time that Rains prepared for President Davis a "memoir upon the use of submarine and subterra shells" which Rains desired to print. The President objected strongly, stating that "no printed paper could be kept secret" and that "your invention would be deprived of a great part of its value if its peculiarities were known to the enemy."[21] Davis did, however, direct that excerpts of Rains' "memoir" be furnished to various commanders for their instruction.[22]

Although Rains had his offices in Richmond, much of the torpedo fabrication work took place in Charleston. Under the direction of Captain M. Martin Gray, 35 to 40 men were employed in their manufacture. Their handiwork was deployed from Georgetown, S.C., to the St. Johns River in Florida, while others were sent as far as Mobile, Alabama. [23] With one boat water torpedoes could be deployed at the rate of four an hour.[24]

As an enthusiastic supporter of Rains' work, Davis ordered him to assist General Joseph Johnston in the defensive efforts around Vicksburg, Mississippi. Rains had little enthusiasm of working for Johnston, as the latter had been very critical of the use of subterra shells early during the Peninsula Campaign the previous year. To ease the way, Davis had Secretary of War James A. Seddon write Johnston expressing the President's desire for the use of Rains' inventions, and adding the War Department "approved and recognized" the devices as approved "legitimate weapons of warfare."[25] Davis told Johnston that "General Rains should now fully apply his invention."[26] Rains' only contribution to the Vicksburg campaign came after the fall of the city and the subsequent evacuation of Jackson on July 16, 1863. On the eve of the evacuation of the city, subterra shells planted along the river bank and on two roads leading from town slowed the pursuing Federal forces.[27]

On August 3, 1863, Johnston ordered Rains to Mobile, Alabama, but President Davis soon intervened and had Rains transferred to Charleston, South Carolina.[28] There Rains, serving under General P.G.T. Beauregard, contributed to the defenses of the harbor and surrounding rivers.[29] While in Charleston Rains witnessed the operation of the submarine *Hunley*, but declined an offer for a demonstration ride.[30]

On February 15, 1864, Rains was ordered back to Mobile,[31] now under the command of Lieutenant General Leonidas Polk, another friend from Rains' cadet days at the Military Academy. Rains directed the mining of the harbor, which subsequently led to the sinking of the Union ironclad *Tecumseh* the following August.[32] After three months in Mobile, Rains was recalled to Richmond. Hunter Davidson, who had overseen the Confederate Navy's torpedo endeavors, was being sent to Europe to work with Maury, and on June 8, 1864, Rains would oversee all such endeavors, taking the "superintendency of all duties of torpedoes."[33]

In addition to subterra shells, submarine mortar batteries, keg torpedoes, and those discharged with electricity, Rains' torpedo bureau also perfected devices which employed a clock-like timer mechanism. As discussed in Appendix 1, John Maxwell placed the most famous of these "horological" mines on a federal ammunition supply ship at City Point, Virginia, causing the explosion of that vessel, two adjacent munitions barges, and a nearby munitions warehouse on August 9, 1864, causing damage then estimated at $4,000,000.[34] A similar horological torpedo destroyed the wharf boat, loaded with munitions and stores for Admiral David D. Porter's fleet, at Mound City, Illinois.[35]

By September 1863, Rains and others began considering the extensive use of subterra shells for the defense of cities, especially to deter raids. As the Confederacy's fortunes continued to decline, the use of Rains' inventions seemed more and more reasonable. Indeed, by the end of the war thousands were in use for the defense of Richmond alone.[36]

As General Philip H. Sheridan's troopers headed towards Union lines on the Virginia peninsula following the fighting at Yellow Tavern, on May 12 they encountered Confederate subterra shells on the Meadow Bridge Road. They sustained several casualties and forced their Confederate prisoners to locate and remove the mines. One Union observer reported "their timid groping and shrinking being a curious and rather entertaining sight." One of the Confederates later reported the work "required a delicate touch and was unpleasantly exciting."[37]

As the grip of the Union forces further tightened on the Confederacy, in mid–October 1864, President Davis ordered Rains to confer with several Richmond battlefront commanders to explain the uses of subterra shells.[38] In addition to mining the roads surrounding Richmond with at least 660 mines by the end of the month,[39] hundreds of mines were placed between Fort Harrison and Chaffin's Bluff, along the north bank of the James River. At General R.E. Lee's request, subterra mines were sent to General John S. Mosby.[40] Rains could now report that "the shells are now appreciated," in contrast to the condemnation he had received in 1862. He noted "the objections to these subterra shells have been met by placing them in lines, with a small red flag, three feet in rear of each, to be removed at night and when threatened with attack, to be restored by daylight."[41]

By the end of November 1864, 1,298 shells had been placed on the road approaches to Richmond. They were being buried at a rate of more than 100 each day.[42] Ultimately 2,363 were placed to protect the Confederate capital.[43] In addition, countless subterra shells were placed between Confederate earthworks and forward barriers.[44] General William T. Sherman encountered Rains' handiwork on the roads approaching Savannah. During the attack on Fort McAllister, on the Ogeechee River, on December 13, 1864. Brigadier General William B. Hazen's Second Division reported torpedoes "blowing many men to atoms" during the attack on the fort; torpedoes caused a majority of the casualties.[45] Subterra mines were used in the defenses of Augusta and Macon. Torpedoes were also incorporated into the defenses of Fort Fisher at the mouth of the Cape Fear River, North Carolina, and in the defense of Battery Wagner on Morris Island, South Carolina.[46]

Rains viewed the subterra shells as "sentinel[s] who never sleep." He was concerned, as were his superiors, about the killing and maiming civilians; however, he also felt they were a legitimate weapon of war which not only served to kill or deter the enemy but also served to demoralize him. Even General William T. Sherman agreed, feeling that torpedoes were "justifiable in war in advance of an army" but when used to blow up trains and roads after they were in Federal possession to be "simply malicious."[47]

At the beginning of 1865, only three major cities remained to the Confederacy — Richmond, Charleston, and Wilmington. By mid–February, Charleston, and then Wilmington, were lost. As the Confederacy collapsed, Rains' agents assisted General Johnston in mining roads around Goldsboro and Kinston, N.C.[48] With the evacuation of Richmond on April, Rains and his family and staff rode with President Davis in the president's railway car to Danville, Virginia; from there he returned to civilian life.

In the years following the war Rains worked as an analytical chemist and was involved with several fertilizer companies. In 1877 he became a clerk with the Quartermaster Department of the United States Army and was stationed in Charleston, South Carolina.[49] He worked on his "Torpedoes" book at least until 1874, judging by articles from *Scientific American* cited in the text.[50] Rains also authored a brief article for the *Southern Historical Society Papers* on torpedoes in 1877 which contains many of the same thoughts, and at times, the same wording, as the manuscript of his *Torpedo Book*.[51] He also wrote about the improved steam safety valve, mentioned in manuscript page 92, which he had patented the year before.[52] The manuscript transcript that follows does not appear to have been the text that Rains gave President Davis in 1863, although the ink drawings may have been from that book.

On August 21, 1874, Rains wrote to President Grant "upon the ground of our old friendship" proposing a course on the "study of Torpedoes at West Point Military Academy." Noting that torpedoes were now used by all "civilized" nations, Rains was "unwilling that a knowledge of this thing, which has been almost the study of a life time, should go elsewhere out of our country." He offered a book containing "some 70 or 80 diagrams with some 150 pages description." These instruments, he felt, would render "war measurably impossible by such formidable inventions." Rains concluded by requesting an interview or referral to scientific gentlemen of Grant's selection. Grant forwarded the letter to Secretary of War William W. Belknap, who on September 24, 1874, endorsed the letter, but wrote that "if the introduction of this study should be contemplated there is no necessity for securing the services of a Rebel General." On October 7, 1874, Colonel Thomas H. Ruger, superintendent of the USMA, endorsed the letter to Belknap: "Instruction in the general subject of torpedoes is now given at the Military Academy including general descriptions of the different kinds, methods of placing, using, &c., which is I think all that should be attempted. To give special instruction with a view to make entire classes of Cadets competent to practically supervise the manufacture of and use of torpedoes would not at the present time be expedient."[53] On October 16, 1874, the War Department notified Rains that his letter had been referred through the War Department to the Superintendent of the USMA and was returned with the Ruger's endorsement noted above, with the additional comment that the "Department concurs."[54]

In 1880 Rains left Charleston and moved to Aiken, South Carolina, where he died on August 6, 1881.[55] Katherine Rains Paddock, the general's daughter, presented the manuscript to The Museum of the Confederacy in 1926.—H.M.S.

Editorial Notes

I have made few changes to the text. A few incorrectly spelled words have been silently corrected, e.g., *Sumter* for *Sumpter* and *Tomb* for *Toombs*. Apparently absent words have been inserted between brackets, e.g. [is]. Where a period has been omitted after an abbreviation, it has been added, e.g., *Genl* replaced by *Genl.* When a word seemed clearly legible but did not appear in an unabridged dictionary, I have spelled it as in the text and added [*sic*]. Words that I could not decipher are indicated as [illegible]; I used this as infrequently as possible. Rains did not use many commas or periods in the text; I have added them silently and sparingly. I have left ampersands in the text as they give some of the original flavor. The few footnotes written by Rains are specifically attributed to him to distinguish them from other editorial notes.

The reader will have to search outside sources for the workings of the Wheatstone electric exploder, Hare's deflagrator, and the Rhumkorff coil used to generate electric sparks for galvanic torpedoes.

The original ink manuscript had no page numbers, they have been added in pencil as two series, a–p followed by 1–108; I have indicated each manuscript page in bold as [**g**], [**63**] within the text. The resulting table of contents follows this introduction, while Rains' differing, original table of contents is included within the work itself. A list of Plates, keyed to the revised manuscript page numbers, follows the contents section preceding Rains' text.

The plates, other than a frontispiece, followed the text. The numbering of some of the plates has been changed within the manuscript at least once; I have reconciled the text and the plate numbering, based on their order in the manuscript. Where the numbers differ, the correct plate number follows in brackets that written in the text. Below each plate I have reproduced the descriptors within the plates, since Rains' writing is almost microscopic in the originals. For ease in following along on the plates, I have inserted them into the appropriate position relative to the text, but I have retained the plate number in the caption. The three plates, and notes on the plates but not referenced in the text appear in Appendix 3.

Rains also had a few line drawings within the text and I have included them as in the original. Four plates have pencil notations below the frame consisting of a series of numbers and letters; their significance is unknown.

The original notebook had thick card covers and was covered with green silk which has largely deteriorated. The text is on blue and cream signatures of ruled paper. The plates are on card stock; those from 22 onward are on various types of card stock differing from the first 21, with the exception of plates 12½ and 14, which are on paper.

The Rains manuscript, including the plates, is the property of the Museum of the Confederacy, Richmond, Virginia, and may not be reproduced without prior written permission.— H.M.S.

Contents

(manuscript page numbers)

Torpedo Book

Gabriel J. Rains
Brigadier General, Confederate States of America

[no manuscript page number] [Table of Contents]

Gabriel Rains' *Torpedo Book*

Defense perfected.—

[The categories are written in ink; the manuscript page numbers were added in pencil, presumably after the text was completed.]

[Pages **a–p** are on a separate signature from the bound text of the *Torpedo Book*, and are located before the bound pages.]

[a] From the *Scientific American* corrected

Remarkable History of a Torpedo Boat from Gen'l [Dabney H.] Maury's report[1]

It was built of boiler iron, about 3.5 feet long and was manned by crew of 9 men, 8 of whom worked the propeller by hand and thigh, steered the boat, and regulated her movements below the surface of the water. She could be submerged at pleasure to any desired depth or could be propelled upon the surface. In smooth, still water her movements were exactly controlled and her speed was about 4 knots. It was intended that she should approach any vessel lying at anchor, pass under her keel, & drag a floating torpedo, which would explode on striking the ship's bottom. She could remain submerged ½ hour with no inconvenience to her crew. Soon after her arrival in Charleston, Lt. [John] Payne, C[onfederate] N[avy] with 8 men volunteered to attack the Federal fleet with her. While preparing for the expedition, the swell of a passing steamer caused the boat to sink suddenly, & all hands, except Lt. Payne, then standing in the open hatchway, perished. She was soon raised, the dead taken out, and again made ready for service. Lt. P. again volunteered to command her. While lying near Fort Sumter, S.C., she capsized, and again sank in deep water drowning all hands except her commander & two others. Bring raised again & prepared for service, Mr. [Horace L.] Hunley (one of her constructors) made an experimental trial to go under a school ship in Charleston harbor, but never came up until found in about 90 feet [of] water with 9 dead men, & brought to the surface, her commander dead with a candle, partially burnt out, in his hand, standing at the hatchway. A forth time prepared. Lieut. [George E.] Dixon of the 21st [Alabama Infantry] Volunteers, with 8 others went out in the harbor on a trial trip, probably, and disappeared perhaps in sinking the steamer *Housatonic* as stated. (As the little boat was found near Ft. Moultrie & not sticking in the *Housatonic* when raised — torpedo A on plate 7 that struck that ship) **[b]** as she had no torpedo at her head she could not sink a vessel by butting. 33 men 1 Negro included were drowned on this boat.

Observations

The boat was brought to the wharf & and left for a long time where we were preparing torpedoes, but on account of the mishaps, I would have nothing to do with her; yet the "thing" is good. She went too deep, in going under the school ship, & perhaps burnt up the oxygen in the atmosphere within as a candle found in the man's hand gives ground for the supposition. When she was found, her head was sticking in the mud at bottom nearby where she had gone down near the ship. With 2 or 3 tanks of compressed air and a plate of iron or lead fitting close to the bottom as a weight to be readily and certainly detached from within to let the boat rise to the surface in case of need; the thing is admissible. [Capt. M. Martin] Gray was the man employed in making & managing the torpedo (plate 7 fig. 1) (which I contrived) and his fears of the result probably induced him to make this supposition of the boat destroying the *Housatonic* as above. That torpedo floating looked like a porpoise and answers to the description given in the northern journals by the men on board that ship who were saved, & not this boat 35 feet long.[2]

[c] [pasted newspaper clippings—not transcribed]

[d] Facts derived from bombardment of Charleston S.C.
Military———Axioms

Long range guns for channel defense should be placed in one or two gun batteries only.

The best of all defenses is an earthwork or sand bags or with natural slope

All exposed masonry should be protected by implants of iron.

Mortar firing if kept up at intervals of certain short duration, will tire out a garrison by depriving of sleep and they will then capitulate after 36 hours disturbance.

Torpedoes for defense of works should be so arranged as not to interfere with night sorties.

The larger the gun the more elevation is required to attain the same range with the same charge of powder.

When sharpshooters are annoying to a work one or more straw men (clothes stuffed with straw) held above the crest of the parapet to receive the shot & moving these about will discourage the marksmen so as to make them careless in their fire. (see remark elsewhere)

Six mortars well served, will stop any advancing sap.

Torpedoes to be fired by electricity are not reliable and are not to be compared with those self acting as in both acting in [manuscript] page 73

Magneto—Electric Fuse
Chlorate of potash
Sub-phosphide of copper
Sub-Sulfide of copper

[e] A strong explosive compound is made by incorporating

chlorate of potash	50 parts
prussiate of potash	23 parts
white sugar	30 [parts]
lead	5 parts

& setting fire by sulfuric acid.[3]

Experience in practical warfare particularly in the defense of Yorktown and the harbor of Charleston, S.C., led to the following suggestions.

All rifled cannon should be well banded with wrought iron, otherwise, they will burst, however, thick they may be.

Elongated shot & shells are the most effective, and there should be at least three calibers in length. Shells fired into a city should have time fuses only.

Eight inch caliber should be the limit of rifle guns, when a heavier grade is used the inertia of the shot either breaks off its rifles, or bursts the gun in turning.

To fire effectively against ironclad rams a smooth bore cannon of fifteen inches is the best weapon, the momentum of such a mass of iron shot is irresistible however guarded with wood within. The Confederacy made a mistake in not adopting it. One at each fort would have checkmated "ironclads."

Floating batteries or flat boats made strong some 60 feet long by 20 in width, with steerage at each end, with a 68[-]pounder rifle pivoted amidships in each, would have been far more effective in offence & defense in Charleston harbor, than the local batteries — for they could not have stopped the ironclad rams, as they were under fire long enough to have passed them & would have passed, but for torpedoes. The batteries to be manned with but 25 or 30 men to be wanted at any moment & to be blown up by subterra shells when taken by the enemy. Two Whitworth guns with hexagonal bore might with advantage take the place of the 68[-]pounder through annoying a fleet a long way off.

[f] The scarp walk of all earth works should be as perpendicular as the nature of ground will permit and to be of such height as to prevent assault except by escalade.

The size of the grain of powder should be increased in proportion to the length of the piece. (Gunpowder can probably be moved with safety in kegs as formerly if mixed with powdered glass, & one of the heads be double so as to press by means of spiral springs upon the surface of the mixture, as motion would soon cause the lighter ingredient, the powder, to come to the top & damaged by being dusted.)

Bombproof vaults should not be permitted to be used except by the sick, as they afford skulkers places in time of action, as at Ft. McAlister, and as safe a place will be found at the breast height of the parapet.

The most effective troops in the world are Infantry.

Breech loaders — from a recumbent position firing and rising to [illegible] at the double quick.

[g]

Fortifications

Impregnable Bastions

The perfection of rifled cannon, their accuracy in aim, and great range, necessitate the withdrawal of all heavy guns from the parapet of bastion works, where they soon would be disabled; and taking them out of view below. The line of sight as at (**a**) plate (21 [*y* in original manuscript]) [manuscript p. 94] some 50 yards from the parapet. This can readily be done by throwing them back on the terreplein of the work sufficiently far (with elevated sights (**bc**) for the shot to clear the parapet, and in its plunge do execution (the dotted line in the plate give the parabola of their trajectory) Smaller caliber mounted on field carriages may still be allowed to give fire over the crest of the parapet as formerly with the proviso that their recoil will also bring them out of sight

To put them <u>in battery</u> after loading, they must needs be forced up their ramp and this necessarily is to be repeated after each discharge. The parapet must be 25 feet thick, and the dimensions of the parts should be as indicated in the drawing. The ditch to be filled with water when it can be done.

There should be a bolster (**d**) to check the piece in a too great recoil and this should be a saucipon[?] or other means as effective for the purpose. The glacis was often omitted or imperfectly made [**h**] yet the abatis was always added when possible and subterra shells (**s**) [**s** is not on fig. 21; the locations of subterra shells are at the tips of the abatis] were found (buried among the trees) as to demoralize the assailants as to render the field work impregnable (for those shells see plate 12 [13] & [manuscript] page 53) Near Fort Harrison on the north side of James River below Richmond there were planted 2,363 of these shells guarding the approaches toward Richmond. No enemy would attempt to remove abatis. Thus strengthened, and as such subterra shell is like a sentinel who never sleeps, the troops behind these report in confidence fancied security so that many men were removed for more active service in the field from these works so guarded. In rear of each shell at 3 feet distance a small red flag (**e**) [plate 21, not labeled; plate 13, **E**] was planted to be removed at night & when necessary and each passageway through the shells, two longer flags [plate 13, **L**] on high staffs (or staves) were planted for ingress, and egress, when required, so at night, two lanterns darkened with red flannel piece hung upon these for the same purpose.

*President [Jefferson] Davis [and] Secretary of War James Seddon had sent for me & asked could so defend these works, which I did.

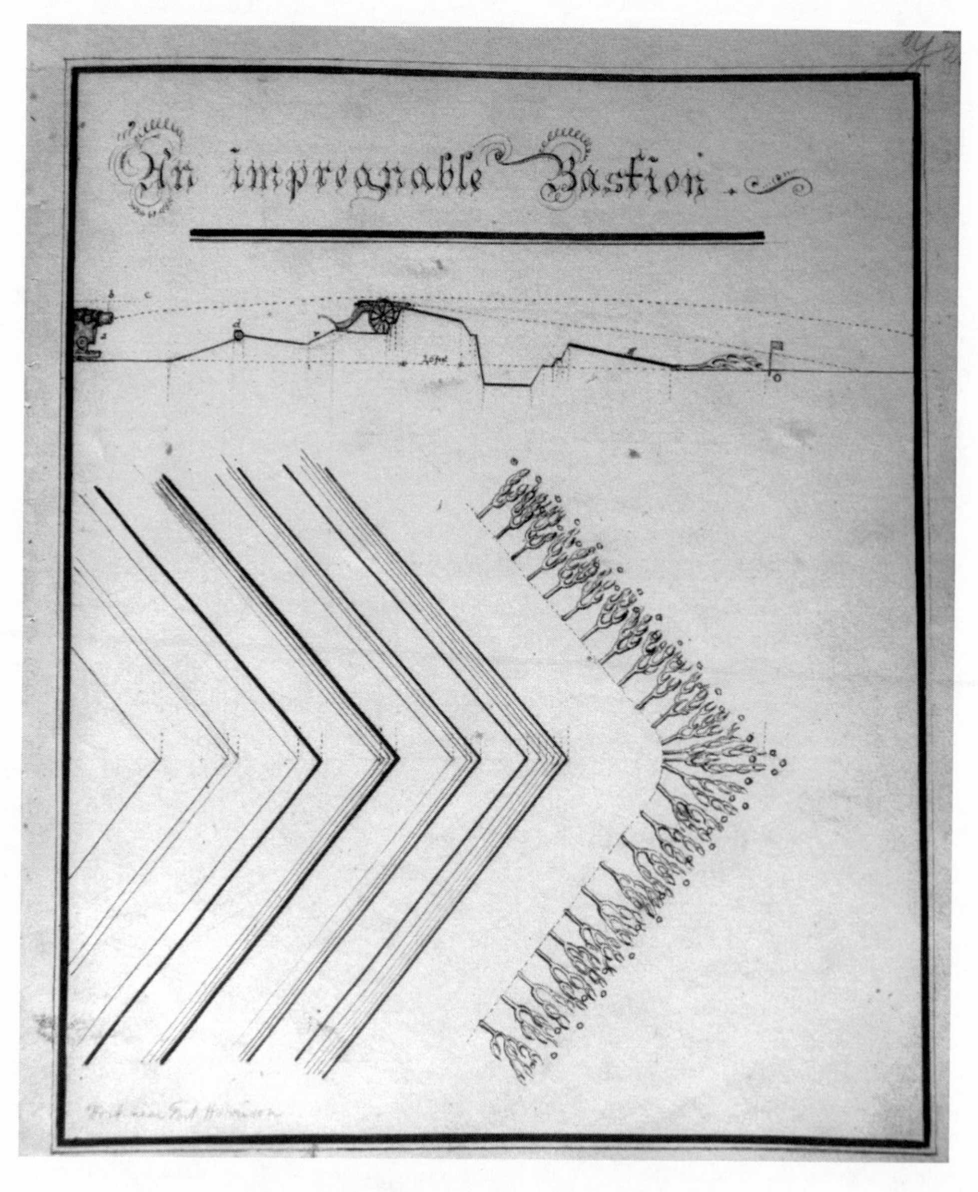

Plate 21: [in pencil left lower corner] Fort near Fort Harrison
[A similar Rains sketch may be found in ORA, Series 1, Vol. 42, part 3, p. 1221.]

[i] To get an electric battery ready

To note an approaching enemy by a sentinel

Intelligence can be communicated by means of a looking glass & flashing iron light a long distance. One flash being (e), two flashes (d or t), three flashes (o) and so on

e	d or t	o	i	s	n	r	a	l	c or k	h	p or b	g or j	u or w
1	2	3	4	5	6	7	8	9	10	11	12	13	14

f or v	m	qu	x or z	[y is missing]
15	16	17	18	

A prisoner from the parapet of a fort where imprisoned (as at Fort Hamilton N[ew] York harbor) might thus unsuspectedly communicate with his friends — say at a time fixed upon, as 10½ o'clock or any other time or to a lookout counting from 3 flashes to excite attention.

This plan of a looking glass is good to lay out a road through woods.

A powerful explosive compound

Chlorate of potash	50 parts
Prussiate of potash	23 parts
White sugar	30 parts
Red lead	5 parts

Another said to be the most powerful of all others

Nitroglycerine	95 "
Colodoin	5 "

Or camphorated gelatin made by adding to nitroglycerine a small percentage of camphor & photographers guncotton previously dissolved in alcohol ether heating gently; when it becomes a pasty yellow cake it keeps well in a warm temperature.

[j] blank page

[k]

Torpedo Boats

The object of this work is to exhibit in detail what has been found useful and meritorious for [illegible] like purposes.

The thousand and one inventions of the many throughout the Confederacy which were subjected to examination of the army by orders of the Dept. of Engineers and War Dept. The commanding Generals of district and in the field, should form no part of it except to note as the abortions of inventive genius. Torpedo boats have attained a kind of celebrity by their short and imposing appearance with their Torpedoes on the ends of beams projecting from their bows & it may be well to state that we have evidence of their usefulness yet fraught with much danger from their proximity to the source of destructiveness *viz.* the exploding torpedoes and probable magazine of the craft attacked (see manuscript page [85]x). The Segar [cigar] boat of Lt. Geo. W. Dixon of Mobile must be also included in this category which was a submarine boat about 35 feet long made somewhat in the shape indicated with capacity sufficient to hold 9 operators. In rear of the boat was her propeller. This boat was constructed of boiler plate iron with a man-hole at top with an iron trap door for ingress and egress having two fins, one on each side near the middle, as steering appendages to be controlled within.

[l] Her crew consisted of 8 men to work the propeller by means of cranks and one to steer. Their speed was about 4 knots. She could remain submerged ½ hour. She was to drag by rope a floating torpedo under a vessel's bottom. This boat could go below the surface & come again to the top but was defective in obtaining proper fresh air for the respiration and sustenance of the crew. She was controllable in smooth but not in rough water. Her efforts in Charleston harbor were attended by towing a Negro — then 5 men in her by being capsized by a steamer's swell when passing with the cover of the man hole off and Lieut. Payne commanding her 9 men in attempting to go under a ship & finally where it is supposed she lies at the bottom and not as falsely reported in the newspapers by going into a hole made by her in striking the *Housatonic* outside of the harbor,[4] where she could not go on account of the roughness of the water and that craft was destroyed by a torpedo noted on (plate 7, fig. 1) which Gray knew, who made statement. The only way to render these torpedo boats effective safely is by detached torpedoes, by butting against the vessel's side with points to stick & hold, then backing off and exploding by a cord, or by pulling one against the ship, or still more efficaciously by carrying torpedoes such as are represented on plates 7 & 8 to be floated by the tides down among the enemy's fleet.

This boat might have been effective by having ½ dozen soda founts filled with compressed air to supply vitiated atmosphere, a barometer tube to tell depth of water & an electric light for a candle.

[**m**] The Russian plan adopted for torpedoes and Lord [Vice-]Admiral [Edmund] Lyons' flag ship blown up at Cronstadt by one, was on the principal of contact and rupture of a glass vial tube containing sulphuric acid commingling with chlorate of potash, sugar & sulphur which readily burns; only gave another evidence of the necessity of immediate contact with vessel's bottom, for torpedoes to be properly available, in destructive effects. It is the secret of my great success with torpedoes. A statement made to Commodore alias Lieutenant [Matthew F.] Maury of nautical celebrity, that I had witnessed 1,500 or 2,500 lbs. [of] gunpowder explode in New York Harbor to disintegrate the rocks, at the bottom, but 30 feet deep, convinced me that the water acted as a cushion to ward off the disruptive effects of the gunpowder, as a boat of 40 tons right over it would not have been injured thereby. (This fact quoted by Com. Maury, which he learned from me, afterward was noted in European papers.) Then torpedoes to be effective must come in contact with the vessels intended to be destroyed & that contact must cause the explosion. This fact was represented to our Engineer Dept. and when acted on accordingly, results have proved the truth of the operation & thoroughly demonstrated the superior efficacy of torpedoes for harbor defense to ironclads and all other means usually employed for this purpose and is now revolutionizing the world subverting other systems of defense, and establishing this as the '*sine quo non*.'

[**n**]

Observations

The almost incompressibility of water at the moment of the explosion of a torpedo under a vessel's bottom must cause the gasses of the gunpowder liberated to act in the direction of least resistance, *viz*, the vessel's bottom, like a great magazine to tear it in pieces, fired from below.

Vessels blown up by these torpedoes containing 60 pounds of powder, the quantity I have directed in my orders, seem to sink immediately like lead.

On an occasion in the James River below Richmond, a gentleman on one of the Confederate boats, a steamer, accidentally striking one of our torpedoes which was sunk so quickly that he made but one step from the pilot house on top, to a boat brought there in a moment.[5]

We have repeated evidences of the fact of the tearing off the whole bottom by these torpedoes, which usually do not allow the crew to escape but they are carried down in the suction.

[o]

Richmond — Charleston — Savannah — Mobile — Wilmington

To burn down a City

Prepare a number of incendiary bullets (see map [*sic*] 15 [16[6]] or 12½) to be carried by small paper balloons over the denoted place to be dropped at night. Of course the time must be taken when the wind is favorable or instead of the weighty bullet, a small rocket case of paper, **R**, filled in its upper part, **T**, with gun cotton paper saturated with spirits of turpentine, and below it gunpowder, **B**, just above and around a glass bulb, **GS**, containing sulphuric acid & below that a plunger, **P**, surrounded by a mixture of chlorate of potash & white vinegar. The machine is to be carried over the denoted place & let fall, when the little weight, **P**, will strike the house or whatever is intended to destroy and the plunger will be immediately driven into the glass bulb, breaking that, when the sulphuric acid will fire the mixture below & then the gunpowder, **B**, which will ignite & blow out the paper, **T**, in a blaze to set fire to its surroundings.

To burn at future time — days intervening — let the plunger be kept stretched by a helical spring and a zinc wire passing into a bottle of muriatic acid & water. When this wire, large or small according to the time, is dissolved by the acid, the plunger becomes liberated to do its work. Common bees wax to cement around the wire at the mouth of the vial with a fine pin hole through it

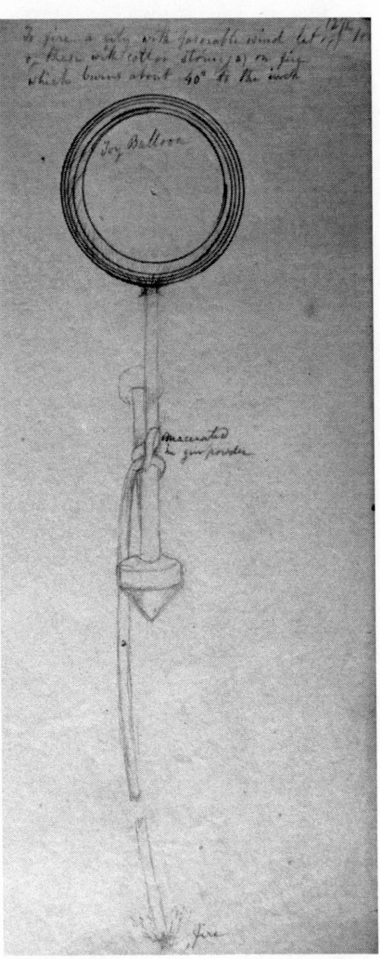

Plate 12½: To fire a city with favorable wind, let off 100 of these with cotton string (a) on fire which burns about 40" [seconds] to the inch

for the escape of the hydrogen gas liberated and to be hung on gimbals in some cases might be necessary.

[p] Torpedo Electric Boat

A cigar boat with a torpedo in front (see plate 18 [19][7]) with no one on board, can be navigated by a wire from shore but rolled on a cylinder there so fixed that positive electricity will turn the rudder one way & negative another way so as to steer. Lay first when about starting, have a lively bed of coals of bituminous or anthracite & a tube leading from a barrel of oil of turpentine or kerosene to drop it on in spray to keep up the fire so as to meet the enemy's vessel. The rudder should be kept straight by a spring until turned to the right or left by the electricity thus regulated.

Let there be a soft horseshoe iron magnet wrapped with coated wire (**m**) and a straight steel magnet above (**a**). Now positive electricity let on will draw down one end of the steel magnet & its torque (**t**) striking one of the surfaces can thus connect with a powerful galvanic battery to pull the tiller of the rudder that way, and negative electricity will repel & draw down the other end of the steel magnet & thus pull the tiller the other way. This torpedo boat with its arm can be made effective better than by ammonia or carbonic acid (see [manuscript] page 97)(& 16) and still better by a fireless cylindrical reservoir containing steam & water heated at 375 lbs pressure per square inch to start with for it may be made to run nearly 6 hours before recharging. See [manuscript] page 107.

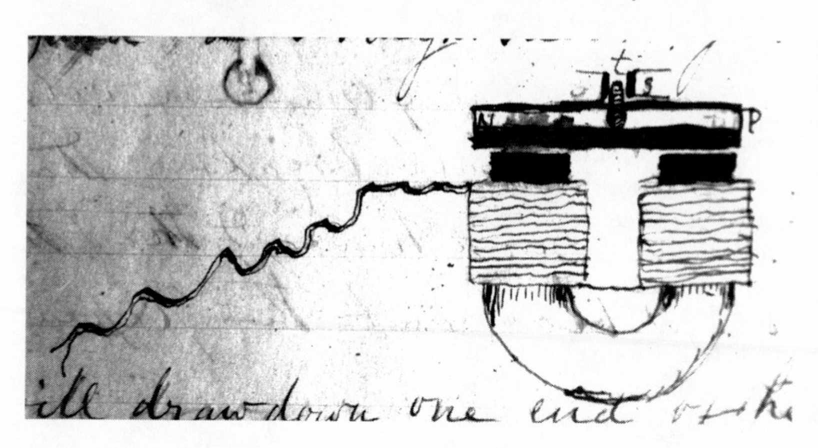

[1] Wooden Torpedo

(plate 1 fig. 1)

To construct these torpedoes procure a number of good strong barrels, beer barrels will answer, and some thick logs, about 15 inches in diameter, from which latter cones are to be made, with bases to fit the tops of the barrels.

By means of nails, join these cones to the heads of the barrels snugly, or instead in each one bore a hole in the apex of each cone perpendicular to the base through the cone and the barrel. Into this pass a screw bolt which is to screw tightly into the barrel head and within, a flat nut must also be screwed down to the head and this can be done and the cone secured in position with another nut at the apex, before the head be put into the barrel (plate 1, fig. 1).

Or better still, procure an iron rod equal in length to both cones and the barrel, and

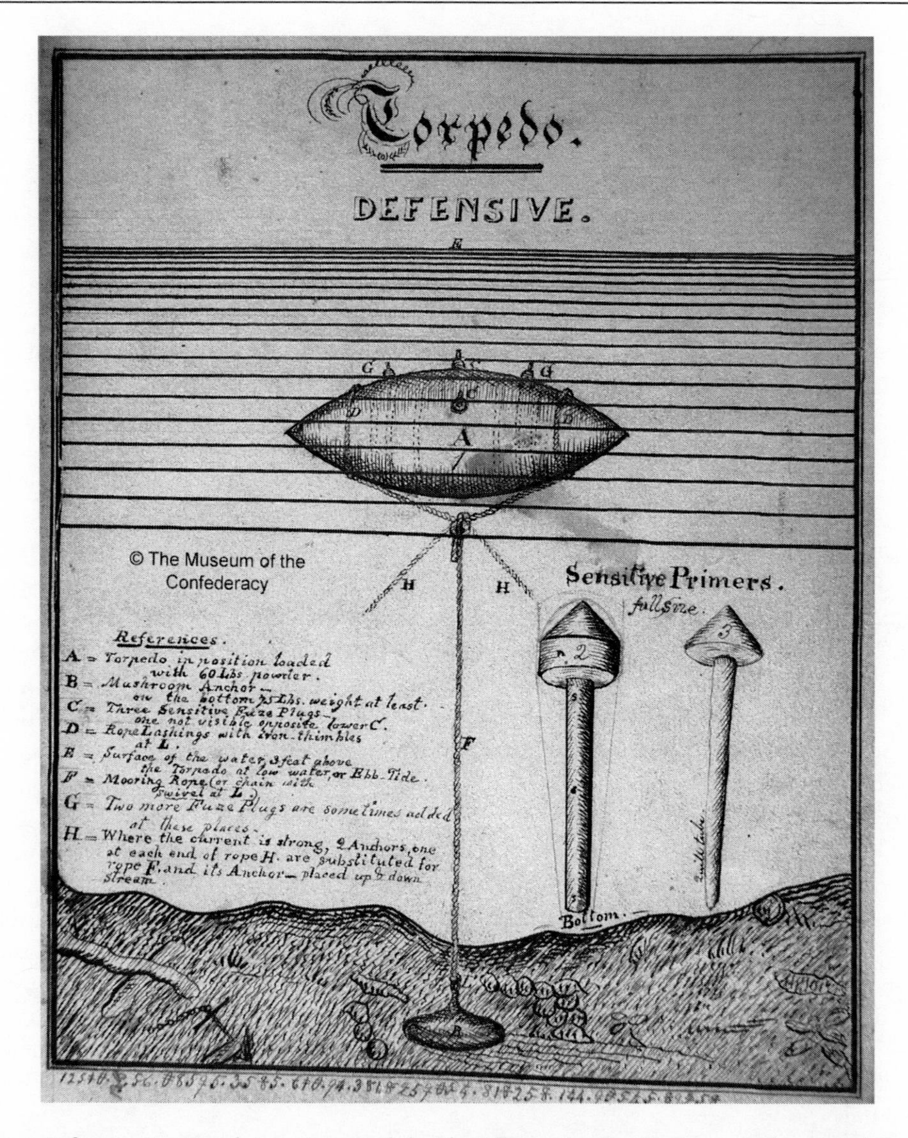

Plate 1: References. ***A***—*Torpedo in position loaded with 60 lbs powder.* ***B***—*Mushroom anchor on the bottom 75 lbs weight at least.* ***C***—*Three sensitive fuse plugs, one not visible opposite lower* ***c'*** ***D***—*Rope lashings with iron thimbles at* ***L*** ***E***—*Surface of the water 3 feet above the torpedo at low water or ebb tide* ***F***—*Mooring rope (or chain with swivel at* ***L****)* ***G***—*Two more fuse plugs are sometimes added* ***H***—*Where the current is strong, 2 anchors, one at each end of rope* ***H*** *are substituted for rope* ***F****, and its anchor—paired up & down stream*

have two screws cut (one at each end) and the rod made in the space between the screws so as to pass tightly through the holes, or have 4 screws cut on it two for the barrel heads and two for the nuts at the cone tops & screw it at once into its place through both cones and barrel.

To prevent the possibility of a leak, better coat the rod with hot pitch & screw in while hot.

[2] After the torpedo is shapened thus the cone head had better be further secured by a few nails and into the bung hole pour hot pitch & turn the torpedo about so as to thoroughly to coat the inside, and then pour out the superfluous pitch.

This pitch was made by dissolving in a kettle on a fire, rosin in coal tar about 4 lbs. to the gallon.

Now coat the outside also with a swab, and same mixture, and after cooling it is ready for the sensitive fuse plug (plate 2 fig. 2) or (plate 5, fig. 2)

The holes for these latter should be now bored so that the screw of the plug should make the female screw, i.e., a little smaller in size, and the prepared sensitive fuse plugs (the helix pitched) screw each into its place. Without the pitching a washer of India rubber might be necessary.

The torpedo bunged up tightly should now be tested for leak by a graduated force pump with gutta percha or flexible hose, and screw pipe, screwed into a hole ½ inch in diameter made for loading with gunpowder. This test is made by plunging the torpedo under water and on applying the air pump observing if any bubbles escapes. Blowing forcibly with the mouth in the hole might answer, but not so certain. Around each end a rope **D** is passed connected as indicated [**3**] with a bellyband which is tied to mooring rope (**F**) made of length sufficient to anchor the torpedo about 3 feet under water at low or ebb tide, by means of the mushroom anchor (**B**).

The torpedo ready for its load, but not put in, may be stored away for future use with the shields upon the sensitive fuse plugs (plate 5, fig. 2) or the protecting wire inserted (plate 2 fig. 2) and this bent to prevent falling out. When wanted for service the gunpowder is put in by means of a funnel in the hole made for this purpose, and a peg driven in the hole to secure it from leak and when the torpedo is put into the water finally, the shields must be unscrewed and taken away or the protecting wires extracted from [it].

The sensitive fuse plug (plate 2 fig. 2) which is made of brass with a leaden disk soldered around it (**l**).

The disk is made of sheet lead very thin such as that of a tea caddy but perfectly water tight. After the plug is cast and finished as indicated, the sensitive primer (**h**) is inserted and then the above disk is soldered on, care being taken in this operation that it is not subjected to too much heat from the soldering iron (copper) to prevent exploding the primer. The wire (**q**) is passed through the plunger and bent and the cover is screwed on after the above operation and the sensitive [**4**] fuse plug then finished, may be kept with others, the same kind, in quantity, until wanted for use. No torpedo, nor torpedo primers, nor primers, nor in fact any thing thereunto belonging, except gunpowder, must be put into a magazine for obvious reasons.

A rectangular prism (**r**) is cast on the lower part of the plug to facilitate its screwing onto a barrel by means of a key made to fit on.

A little beeswax is forced into the hole (**o**) (so as to close it) with the finger, and some white lead put up the helix of the screw (**p**) so that when screwed in, it may make a water tight joint.

A vessel passing over a torpedo must necessarily come in contact, at the bottom, with the plunger, and push that down upon the thin metallic disk, which yielding in its turn will explode the primer & torpedo. Experience has not proved which of the two forms of sensitive fuse plugs, this, or that on plate 5, fig. 2, is the best. The essential difference of the latter being a oval copper cap (**G**) soldered on to be impinged upon & pressed in by the bottom of the vessel or ironclad, as she passes over it. The shield having been removed for that purpose.

For salt water, where barnacles & small oysters might grow upon the plunger & impede its action undoubtedly then [**5**] that with the copper oval cap is best, care being taken the

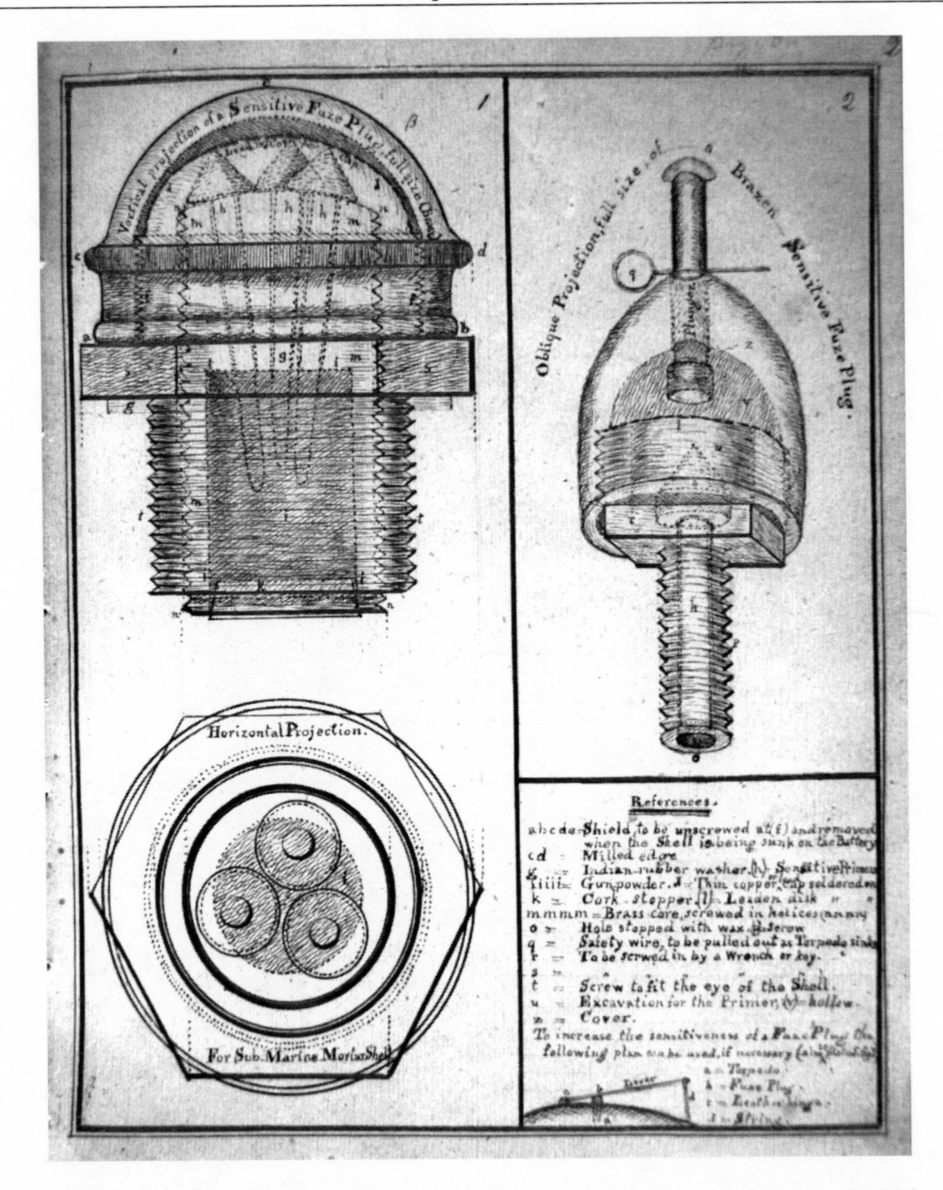

Plate 2: References: ***abcde****—shield to be unscrewed at (f) and removed when the shell is sunk on the battery* ***cd****—milled edge* ***g****—Indian rubber washer (****h****) sensitive primer* ***iiiii****—gunpowder* ***d****—thin copper or lead soldered on* ***k****—cork stopper* ***l****—leaden disk soldered on* ***mmmm****—brass core, screwed in helices (****nnnn****)* ***o****—hole stopped with wax (****y****)—screw* ***q****—safety wire to be pulled out as torpedo sinks* ***r****—to be screwed in by wrench or key* ***s****—to be screwed in by wrench or key* ***t****—screw to fit the eye of the shell* ***u****—excavation for the primer (****v****) hollow* ***x****—cover*

To increase the sensitivity of the fuse plugs the following plan can be used if necessary (also see plate 5, fig. 2) ***a****—torpedo* ***b****—fuse plug* ***c****—leather hinge* ***d****—string*

copper be so thin, as to give way readily from pressure of the thumb or finger. A thin lead cap was substituted for the copper one, and many sensitive fuse plugs thus made, but we have no evidence of their superiority. Thin copper, well annealed, answers all purposes.

The base of that in fig. 2, plate 5, being a hexagonal prism require a corresponding key to screw it in.

Observations

In all cases the torpedo with its fuse plugs, *&c*, *&c*, must be thoroughly made, and finished, before its load of gunpowder put in.

The vacancies around the sensitive primers are usually filled with fine gunpowder, except where it might interfere with the pressure thereon.

In filling excavations (**i**) plate 2 fig.1—fine powder is just next to the primers, and then coarse grained or canon powder is put in, and the cork stopper (**k**) leaving a little space so as [to] prevent the possibility of any undue pressure setting off the primers.

The wooden torpedo will last for months perfectly water tight & is easily made with the material (beer barrels) always at hand provided you have the sensitive fuse plugs at command.

In an emergency a torpedo can be constructed in [**6**] an hour by driving a staple into the head of a beer barrel for an anchor rope & in the center of the other, & on the "sides" boring holes & screwing in sensitive fuse plugs—filling with gunpowder and anchoring with a common boat's anchor if so required. Of course this is but a temporary substitute.

For the harbor of Charleston, S.C., & in Stono River there were 123 torpedoes planted mostly of the kind in plate 1 and also near as many in Mobile harbor, and in other places, where they were found effective after they had drifted loose in part upon the beach, months after they were planted. But some with a modification of fuse done by Capt. Gray were certainly defective as they (plate 2nd fig. 2) had the plunger to come down upon the primer, the leak around the plunger being prevented by stuffing. To remedy this I had interposed a thin lead disk soldered on so as to be impervious to water, and made the plunger loose to be pushed down when hit by contact. A number of these defective torpedoes in Stono River & Charleston harbor, Capt. Gray made in my absence, who though a good workman was no inventor.

[7]

To construct the Submarine Mortar Battery Plate 3

Procure 3 pine logs some 60 feet long which may be hewed or not as fancy dictates, but must be beveled at the larger end, at an angle of about 45°, which must admit of an area 1 foot square marked **BBB** in the diagram to put submarine mortar shells upon. In about 6 feet from the end thus prepared bolt on 4 pieces of light seasoned timber about 35 feet long, so as to allow 15 feet between each beam, these logs are marked **CCCC** and the bolts E (the floats must be of light wood). Near the foot of the beams **BBB** bolt a tie piece **D** to each, and drive in the 3 staples so as to secure thereby the six 10 inch shells, used as weights or sinkers. Each shell will weigh about 112 lbs. and these can be fastened by screw links in the eye, they are marked **NNN**.

A spike or pointed iron pin is to be driven into the small end of each beam marked **I**.

Through a hole at **K** in the center beam near the place of the shell, pass the chain **H**, having a "T" at the end **I** and is fastened on to the weight or mushroom anchor **J**.

Fasten one end of the wire **G** (such as is used for telegraphs) on bolt E, and the other end into spike **q** about the same bolt, or around the beam fasten the buoy rope and buoy.

The ends of the floats **CCCC** must be tarred up or pitched so as to prevent water soaking in, which [**8**] would soon render them water logged.

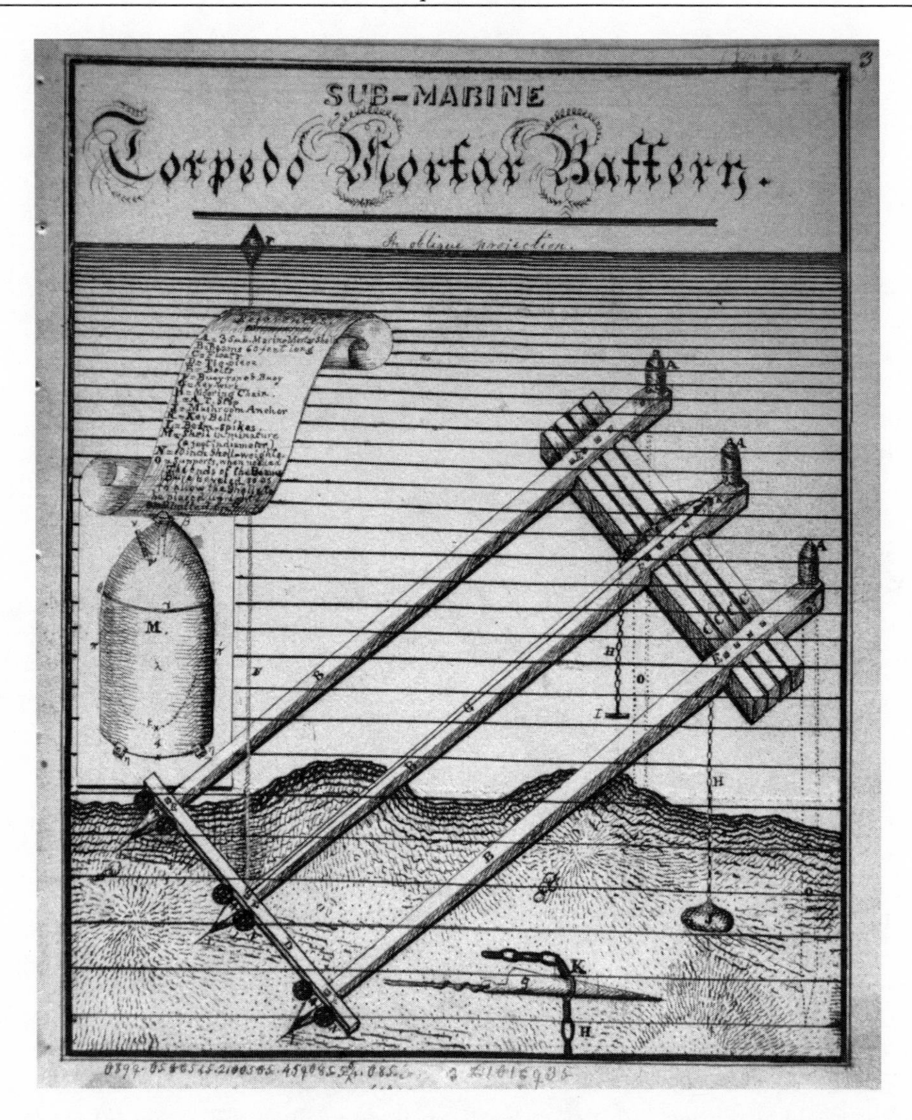

Plate 3: ***A****— 3 submarine mortar shells* ***B****— Beam 60 feet long* ***C****— Floats* ***D****—Tie-pieces* ***E****— bolts* ***F****— buoy rope & buoy* ***G****— key-wire* ***H****— mooring chain* ***I****— A T stop* ***J****— mushroom anchor* ***K****— key bolt* ***L****— beam spikes* ***M****— shell in miniature (a foot in diameter)* ***N****—10 inch shell-weights* ***O****— supports, when needed The ends of the beams (****B****) are beveled so as to allow the shell to be placed upright and bolted on.*

To provide for such a contingency there is usually bolted on near each shell the pieces (**O**) marked, which are weighted at the lower end if necessary. They are placed lengthwise with the beams when the battery is moving to its destination. Finally, there is to be bolted on the

Submarine Mortar Shells[8]

This shell marked in the diagram M is cast with a hollow something of the shape of a lower mortar chamber or parabolical, the vertex at (**γ**) where it is 4 inches thick. It is from

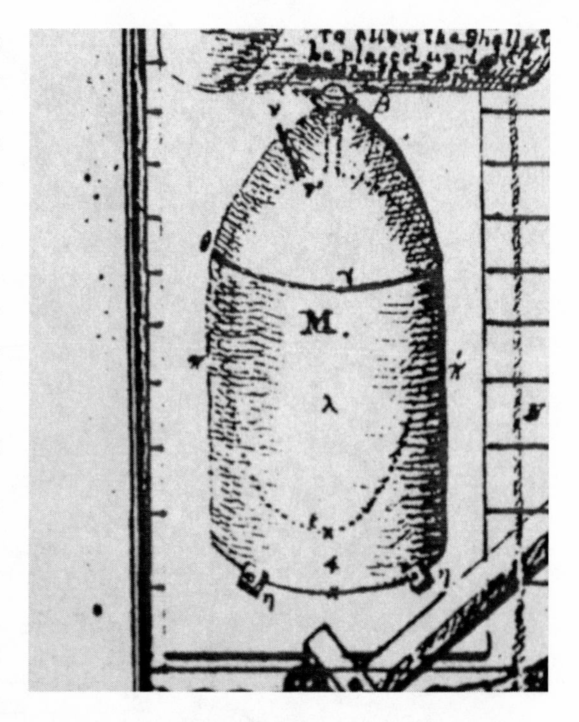

(**π**) to (**π'**)1 foot in diameter outside & is 2 feet long having flanges at (**η**) 4 in number with holes for bolts in each. The interior declines in thickness to (**θ**) where it is an inch thick only & is still more reduced by the furrow (**γ**) which is turned into it. From this, it increases in thickness, becoming hemispheric in shape within, & parabolical outside, to the fuse hole or eye, which is 2¼ inches in diameter, cut into a female screw, so as to admit the sensitive fuse plug (**β**) screwed in. At (**γσγ'**) is a hole ½ inch in diameter stopped with a tapering screw tap at (**γ**).

This hole is intended for loading the shell at last moment when fixed in position at (**A**) and about to be towed to its place of destination. When the shell is fired it is intended to part at the furrow (**θσγ**), the lower part (**λ**) giving off the upper part into the vessel's bottom, as a mortar would [**9**] a shell. The load is about 25 lbs. of gunpowder. The sensitive fuse plug made of brass has a shield **abcde** (plate 2, fig. 1) which is to [be] unscrewed and then there will be exposed a lead or copper disk or oval **ddd** made of very thin copper or lead, which can be pressed in, with the thumb, and such as when impinged upon by a vessel's bottom, it will readily yield upon the primers (**h**) and explode them. (plate 2, fig. 1).

The sensitive primers are inserted [into] holes bored in core (**mmmm**) at top, and then the core screwed into its place as indicated in the drawing, and then the thread of the screw [is] nicked to prevent it turning either way. After the primers are inserted and fixed, the core screwed in, and nicked, the space (**iiii**) is to be nearly filled with gunpowder. First give them coarse grained and the cork stopper (**k**) is to be put in.

The plug is now ready for the shell, and putting some white lead on the threads of the screw, it should be screwed into the top of the shell, down upon the washer (**g**) so as to be perfectly water tight (the loading being deferred until needed by way of safety).

The fuse plug is to be screwed in the eye of the shell by means of a key made to fit the hexagonal prism (**sss**), and the shell receive[s] its powder (about 25 lbs) by means of a small funnel inserted in hole (**γ,γ'**) (fig. 3).

To remove the Battery

Suppose it planted as indicated, where nothing is seen on the surface of the water but the buoy [**10**] which may be a fishing cork.

With 2 boats (yawls if convenient) with 5 men each (4 to row & 1 to steer) approach the buoy, seize the rope, then pull it over the rowlock or "scullock" in rear, until the foot of the battery is brought to the surface of the water. There 'belay' it with another short piece of rope so as to be easily let go. Now seize the wire (**G**) & pull it out from the beam when it will draw out the pin (**q**) from the link of chain at (**K**) when the anchor chain **HH** will run out. The (**T**) and the shells come to the surface. With a hook pole catch the chain below the beam and raise the mushroom anchor (**J**) into the other boat & row with the first boat, the other leading, to the place [to] set it anew. It will require about an hour to remove the battery & but a moment to plant it.

On arriving at the spot to locate it, at a call or signal the 1st boat lets go the foot to sink to the bottom & the other boat's crew throws over the anchor having made the chain **H** the proper length from the mushroom anchor to beam and hole (**K**) by means of the key (**q**).

The buoy **F** is to be small as possible or a common quart bottle will suffice, so as to avoid exciting attention — otherwise the Buoy must be removed, and a grapnel resorted to for future removal of the battery.

To load the shells, use the finer grain powder, this remark applies to all torpedoes. The coarse grain will answer, but is not so instantaneous and effective.

[11] Observations

Torpedoes, floating boats, &*c.* should be colored black also the faces of men & their clothing, hands, &c.

Submarine mortar battery is useful to close a channel, where the enemy is likely in dark nights to use their small boats for the removal of torpedoes. They are too heavy to be removed by dragging, and their effectiveness was proved at Charleston, S.C., early by a cotton guarded steamer which negligently was permitted to float down upon one, which sank her in less than a minute.[9] Two of the shells were exploded. Flat boats were used effectively to test their merits in the beginning, but since, they have done service & probably by saving Charleston proved their merits as Admiral [John A.] Dahlgren reported. The torpedoes in the harbor there of a nature that they could not remove them. There were in the Stono River and in the harbor of that city some 25 torpedoes of this character.

As the submarine mortar shell is impervious to water [it] will last for years, and with the shield of the destructive fuse plug on, they may be made useful in many ways, as they are then safe & can be handled with simplicity.

They are likely to become waterlogged after a great length of time immersed in deep water, for which the end pieces were affixed — yet in Charleston Harbor we had no evidence of this fact, after due inspections, yet should it occur, the pieces as above will still render the battery effective until destroyed by worms.

[12] Gun Cotton

To fill a torpedo with gun cotton in many cases is desirable and in shells it is more effective than gun powder, yet though more than twice as strong, it require[s] more than twice the open space to occupy, and the difference in cost is mostly in favor of powder.

Gun cotton is a true chemical compound and will keep any length of time without deteriorating, but to fill shells, it had better be warmed to ensure its dryness just before being used.

Quick match should always be made of gun cotton if possible; it is a great improvement, as it fires at a much lower temperature than gunpowder & when somewhat macerated in meal [i.e., finely ground] powder, it readily communicates fire to that article. Hence for the fuses or shells, &*c*., it is very superior.

Observation

On an occasion, a noise was heard at night in the harbor of Charleston and one or more dull sounds succeeded, but daylight only revealed the mystery. For it seems the enemy attempted to move in their ironclads, which coming in contact with these batteries, sunk one or two of them as their smokestacks indicated, sticking out of water next morning. The enemy acknowledges one, *Weehawken,* but the torpedo (see plate 3) was not put there as supposed especially for her, for it had been put there long before to close the channel. This came up nightly & threw shells into Charleston at midnight among women & children & was lucky in missing the submarine monitor battery until December, 1863, when she struck it.[10]

[13] Circumvention of the Devil Machine

There came into the harbor of Charleston a machine ingenuously contrived & fitting on the bow of a steamer to take up or neutralize the effect of the torpedoes. There were said to be several of these but the one in plate 4 was well calculated to succeed which led to the invention of the circumventor which is thus made.

Take an ordinary torpedo — in the present instance that described on plate 1, and on the upper part screw in the 5 sensitive fuse plugs as indicated in plate 4, fig. 3 and attach it to a line, circled up by a cord knotted as shown.

This cord pulls out on breaking the cotton string at (E) and on arriving at the coil and winch as (A) it lets the coil loose, and the torpedo immediately rises towards the surface in the position indicated in fig. 2.

In figure 1 the circumventor is shown planted in position with a block of wood at the end of the line (C) floating about 3 feet from the surface of the water at ebb tide, and the anchor upon the bottom heavy & with flukes to prevent being carried away. The Devil is in the very act of seizing the block of wood (or this might be a torpedo if necessary) and on stretching the line (C) the bottom tie at (E) will give way [14] and the line will begin to pull out the loops at (D).

After these are all pulled out, and the steamer having the Devil in front passes over the torpedo that is free to rise up against her bottom and destroy her.

The lower part of block (B) is made slightly conical, so that should attempts be made by small boats, any rope sunk below the surface to sweep the place, it will pass over the block, which will yield before it and remain as before.

For this block of wood a torpedo (as in plate 5 [4] fig. 1) can be substituted so as to be effective, should either attempt be made to remove, or search made for them.

The cord in [plate 4,] fig. 3 is made slack so as to be understock [*sic*], but it is to be pulled tight at the top of the anchor (A) so as to secure the coil (F) against it when used.

The machine with a block of wood at top was planted in Charleston Harbor but with what results is unknown except that it destroyed the *Housatonic* probably.[11]

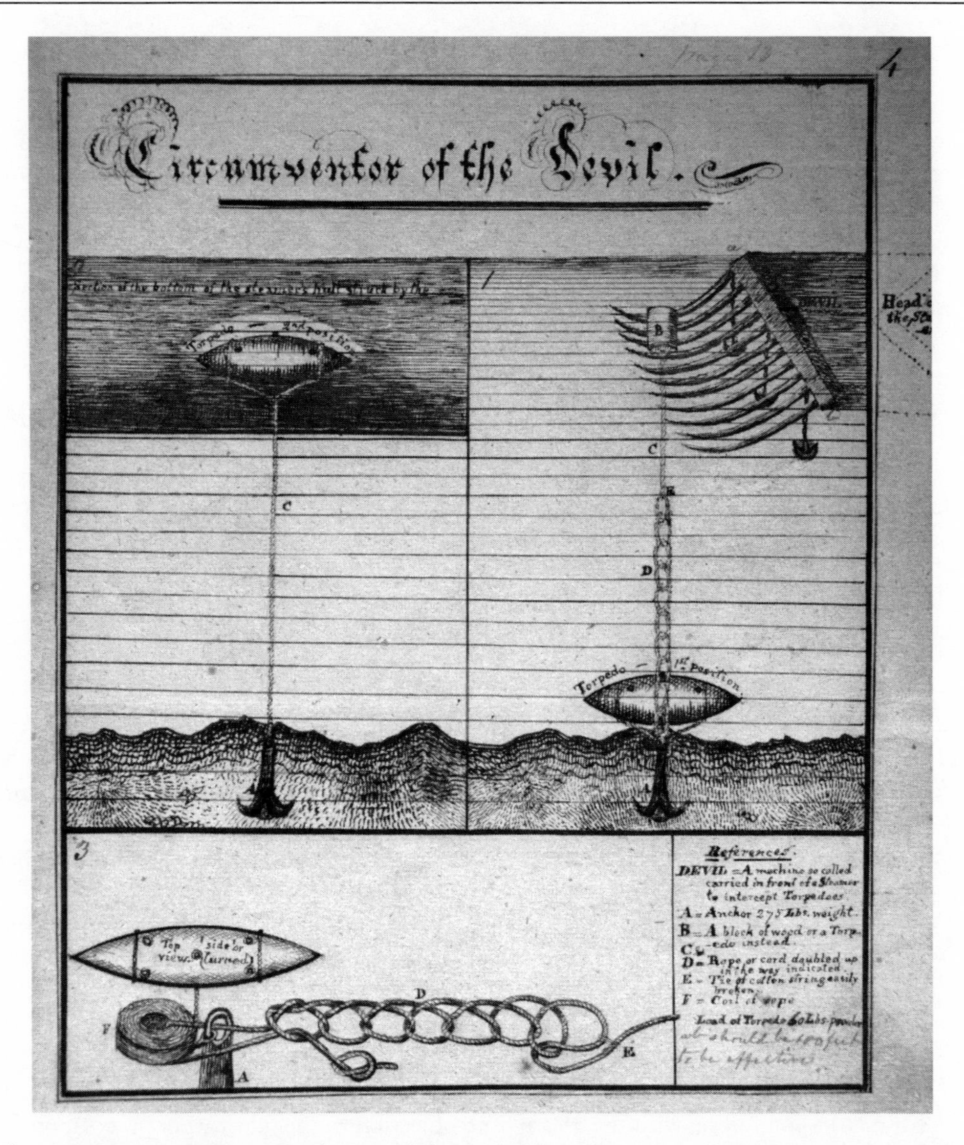

Plate 4: References: ***DEVIL****—A machine so called carried in front of a steamer to intercept torpedoes* ***A****—anchor 275 lbs. weight* ***B****—a block of wood or a torpedo instead* ***C—D****—rope or cord doubled up in the way indicated* ***E****—tie or cotton string easily broken* ***F****—coil or rope*

Load the torpedo 60 lbs powder [in pencil] ***ab*** *should be 100 feet to be effective.*

The Devils were not used, or were found ineffective, after the first attempt of the circumventor. One of the Devils was obtained and exhibited in Charleston, S.C.

Observations

The coil of this torpedo can be at shore, and the rope from it extended along the bottom of the anchor & there passing through a block or roller attached with a swindle or swivel can keep the torpedo down or be used to pull it down and let friendly vessels pass to be let out in time for the enemy's ships.

[15] Observations

Bottles with papers containing information for the enemy were sent down from Charleston by the tide, to be picked up by them, some of which were intercepted by Confederate boats. It would have been easy to have made such into traps to explode with gunpowder by means of a sensitive primer on withdrawing the cork. It was not done, however, the whole object being rather to demoralize than kill and to act on the masses not individuals, so for torpedo purposes the largest shells were selected usually for subterra shells, to be noted hereafter, and torpedoes filled with 60 lbs of powder, that's what was done should be well done. A proper and judicious size of torpedo should ever be borne in mind, and this tremendous engine of destruction should never be used where innocents would suffer.

Maj. Gen. [Leonidas] Polk ordered a man to report to me who was desirous to blow up rail road cars. I refused to accord with his scheme except were the train was used for soldiers, or military supplies only. Human life should not be trifled with, yet to frustrate the attempts of a midnight assassin, or of an enemy invading your soil & seeking your death, their destruction cannot be against the laws of God, when self-preservation is the law of nature.

[16] Torpedo Boat

A boat for protection of harbors may be made of iron plates — 30 feet long 3 feet wide — propeller-deck nearly level with the water — cigar shaped to be managed by three electric wires insulated, in one cable wound on a drum on shore to control the steerage & throttle valve, the propelling force ammonia liquefied, which at 60° Fahrenheit gives 60 lbs. pressure to the square inch, to be applied like steam, with two vibrating cylinders.

For service the ammonia is poured into a system of tubes located in a water tank so that the gas which passes through the cylinders is carried by the exhaust pipe to be absorbed by the water, so as to economize being used again and again over a thousand times if necessary.

This boat to carry on its stem one, two or three projecting beams from the bow with torpedoes, one on each beam's end which should be same 10 to 20 feet in front of the boat. The kind of torpedo preferred should be like (**M**) in plate 3, only of copper or tin so as to be light.

As ammonia acts on copper & brass, the cylinders & their packing & mounting should be of iron & its combining with the grease is no objection to the lubrication in its saponification[?]. The throttle valve might be a simple stop-cock. At 40° below zero the boiling point of ammonia, it has 14 lbs pressure the same as water at 212° and if we add 40° below to 60° above we will get 100 lbs pressure to the [square] inch and a bottle of liquid ammonia which one can carry in his pocket would be sufficient to drive an enemy out of the harbor or to "Davy Jones' Locker." See Dr. Emile Lamm's invention of ammonia engine narrated in *Scientific American*, Vol. XXV, page 290.[12]

[17] Vertical Wood Torpedo (plate 5, fig. 1)

This torpedo like others of wood already discussed, is make of a barrel pitched both within & without. The bolt (**B**) passes through the head of the barrel where it is secured with a nut (**a**).

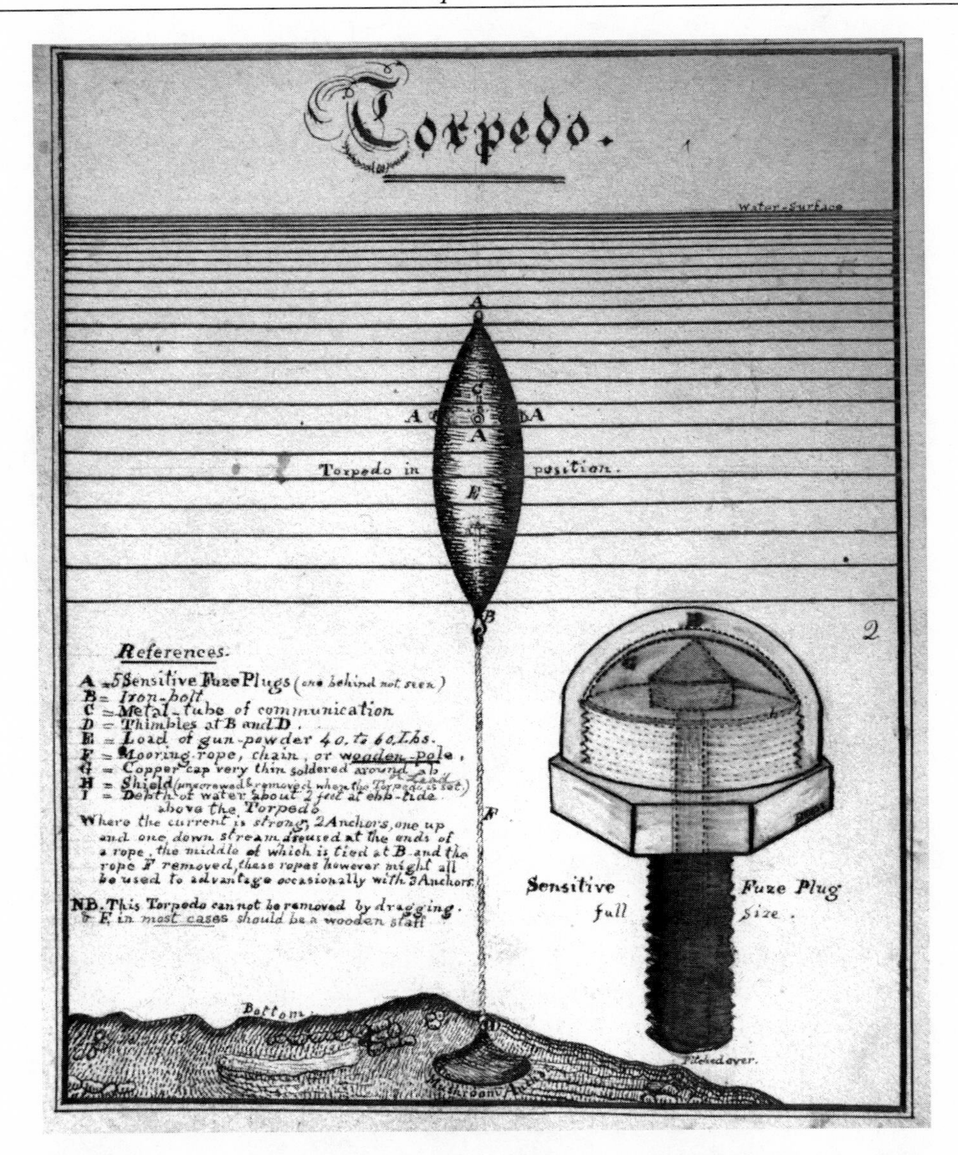

Plate 5: References: ***A****— 5 sensitive fuse plugs (one behind not seen)* ***B****— iron bolt* ***C****— Metal tube of communication* ***D****— thimbles at* ***B*** *and* ***D E****— load of gun powder 40 to 60 lbs.* ***F****— mooring rope, chain, or wooden pole* ***G****— copper cap very thin soldered around* ***ab*** *or lead* ***H****— shield (unscrewed & removed when the torpedo is set* ***I****— depth of water about 2 feet at ebb-tide above torpedo.*

Where the current is strong, 2 anchors, one up and one down stream are used at the ends of a rope, the middle of which is tied at ***B*** *and the rope* ***F*** *removed, these ropes however might all be used to advantage occasionally with 3 anchors.*

NB. This torpedo cannot be removed by dragging & ***F*** *in <u>most cases</u> should be a wooden staff.*

In the top at (**A**) the sensitive fuse plug fits into a tube (**C**) made of metal, copper or lead, to which it is soldered water tight. The lower end of this tube ending in a screw by which it is screwed into the head of the barrel or keg (½ barrel).

Around the torpedo just below the chime of the barrel, some 5 sensitive fuse plugs are screwed in, and where the bolt terminates as (**B**) an iron thimble is used to prevent the rope chafing against the iron which should soon wear it apart.

The mooring rope is made of such length as to bring the torpedo in about 3 feet of the surface at ebb tide, counting from the apex (**A**). This rope must also have a thimble at (**D**) next to the anchor.

As these torpedoes explode only on contact, there is less powder required than those placed at the bottom of the water, as from observation of large quantities of gunpowder thus exploded in New York harbor. The water acts as a cushion to prevent the ill effects of gunpowder, a fact to be noted as a 40-ton boat moored over 2,500 lbs powder 40 feet deep would not be injured in explosion.
[**18**] (This was seen quoted in some European papers as the remark of another (Lt. [Matthew F.] Maury of naval celebrity) to them communicated it as my own observation.[)]

But how small the quantity of gunpowder should be to destroy a vessel has not been determined exactly. The sudden development of gasses from exploding gunpowder under the bottom of a vessel, on account of the measurable incompressibility of water, must impinge against that, comparatively as a solid, and the weakest direction of escape must be through the vessel's bottom.

The great fault of the Russian torpedoes at Cronstadt was their having too little powder. Through ignorance of the fact noted and probably were on the bottom as Lord [Vice]-Admiral [Edmund] Lyons' flag ship was raised partly out of the water by one — yet afterward was found not to have a plank started. Sixty pounds of gunpowder tears the vessel's bottom all to pieces, so that she sinks in less than a minute & thus carries down the crew with her but very few making escape. Many that could get off must be taken in by the suction, which necessarily follows from the sinking of a large and heavy body.

Twenty-five pounds of powder will blow up a house of 4 rooms which lifts the roof up & bursts out the sides, but this must have the explosion of the air within to aid its effects.

For the case of torpedoes no such calculation can be made acting as it must against a strong copper or iron sheathed structure so the quantity had better [**19**] be increased, in fact had better be too much than too little, so not less then 40 lbs. of gunpowder should be thus employed, and this is probably insufficient for all torpedoes, which explode by contact (thus we use 60 lbs.)

The vertical torpedo cannot be easily removed by small boats dragging in the night time, as the drag rope when it comes in coalition with it, passes over that and its fuse plugs, without exploding them, the torpedo yielding before it.

The location of each torpedo should be noted,[13] and the spot marked on a map of the harbor prepared for this purpose, so that when necessary, they can readily be found and taken up, as I had done in Charleston.

The horizontal torpedo can easily be removed & examined by a drag line, but this vertical one is much more difficult and ordinarily its removal is impossible except by prepared means, its locality known. Instead of a rope a chain might be used to anchor the torpedo, with swivel between (**B**) and (**D**) and such was found most efficient in Mobile Bay. The chain the size of the ordinary trace chain, and if galvanized the better. The chain will not fret apart nor be rotted by salt water. Another advantage in the chain, it cannot be cut as rope might be under water, to release the torpedo.

These torpedoes are apt to lean with the tide and a strong current might enable a vessel to pass over [**20**] them harmless, yet where this is liable to be the case, two anchors connected with a rope should be sunk, one up and one down stream, distant apart, so that the center of the rope brought near the surface of the water, there the torpedo is to be fastened, it shall form an equilateral triangle. This will keep the torpedo in position at all times whether the current be running out or in.

To remove the vertical torpedo (plate 5 fig. 1) a boat with one end of a line must be stationary, and this lead line sunk, must be carried around by another boat to make a coil about the anchoring rope to raise it. The small boats might go around each in the opposite direction passing one another, & then rowed off carrying the torpedo with them.

None but small shallow boats should be used to take up torpedoes, as these do not draw so deep in water as to make contact, & when so shallow they can be easily seen.

Observations

Through carelessness a shallow boat with a Lieutenant C[onfederate] S[tates] & crew of 2 or 3 men did come in contact with a torpedo in Charleston harbor. It blew off the end of his boat & they had to swim for it, but none were killed. It was ebb tide at the time, and the crew was negligent in not looking out and careless in going over the torpedoes.

[21]

The Tin Torpedo (plate 6, fig. 1)

is easily constructed of sheet tin by any ordinary tinner. It is shapened as indicated in the figure with an oval top or conical, slightly so raised as to strengthen it to resist the pressure of the water.

Onto this top is soldered 5 sensitive fuse plugs and a small hole at (f) for the purpose of loading is made to be covered finally with a disk of tin by soldering. This is better than a cork put into a short cylinder & cemented over usually done, though this latter will answer for small depths & is much more convenient and safe.

At the bottom there is a tin socket (C) for the staff (D) to fit into, where it is secured by a rivet or nails. This torpedo is connected to its mushroom anchor by a wooden staff made of the proper length to suit the depth required.

This plan has some advantages, as a boat coming in contact with the top, the vertical pole will resist the impact and force pressure upon the sensitive fuse plugs to explode them. It can be secured in position by the means indicated, *viz.*, two anchors up & and down stream and a rope.

The joint at (E) might be a socket instead, and then the buoyancy of the torpedo by means of air in the upper part above the gunpowder will give and [22] keep it in position.

This is a very effective torpedo, and was much used in the James River below Richmond. The charge usually was from 40 to about 50 pounds of powder. The tin torpedoes before loading can be easily tested to see if they leak by sinking them in barrel of water and sucking with the mouth at (f). After the operation, should they be faulty, water will be found within, as seen through the hole (f) the bright tin reflecting the light.

As a further precaution after being thus examined they should be painted over to ensure protection against salt water.

As the pressure of water increases as the square of the depth, tin torpedoes must be so

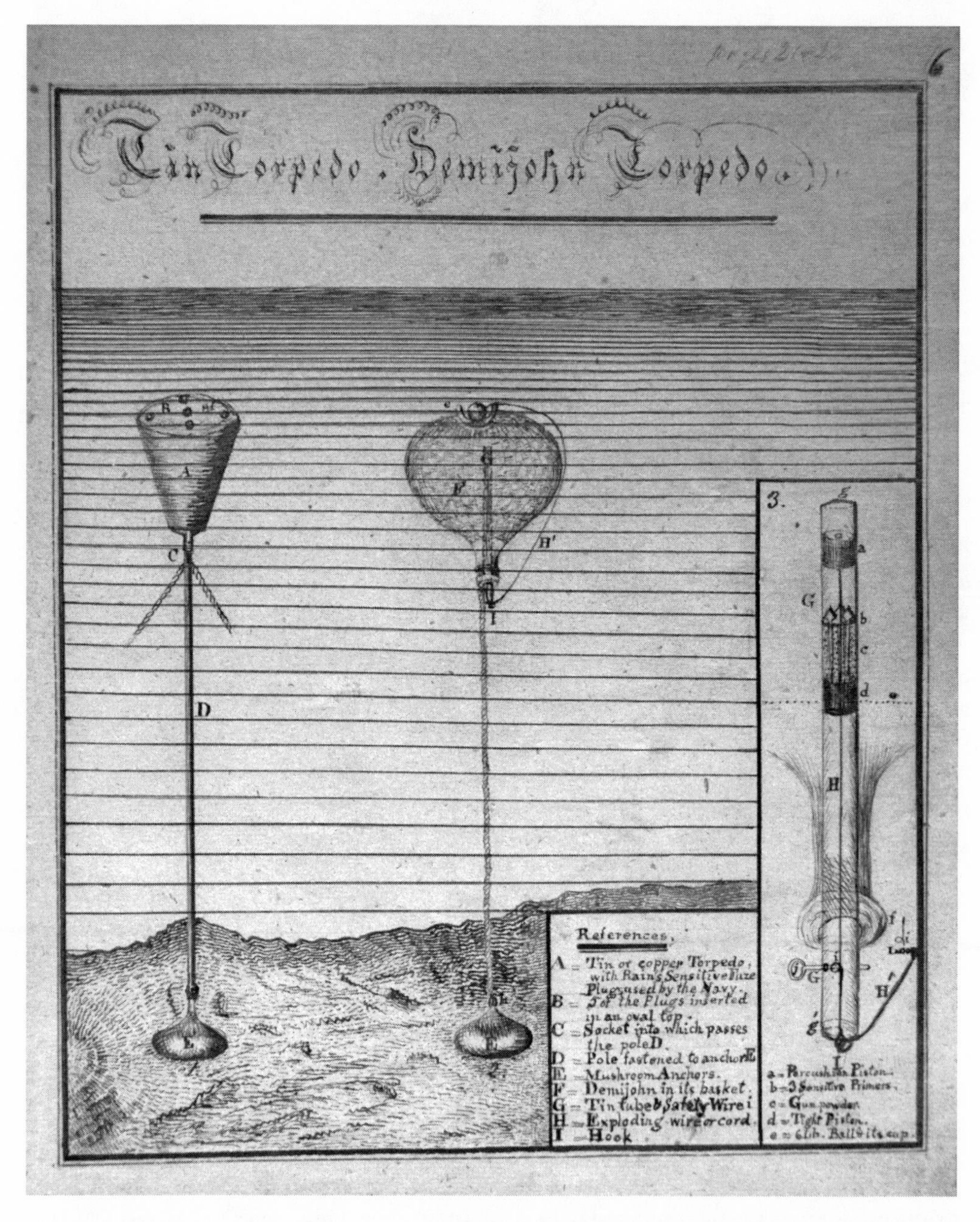

Plate 6: References: ***A****— tin or copper torpedo, with Rains sensitive plug used by the navy* ***B****— 5 of the plugs inserted in an oval top* ***C****— sock into which passed the pole* ***E D****— pole fastened to anchor* ***E E****— mushroom anchors* ***F****— demijohn in its basket* ***G****— tin tube* ***b****, safety wire* ***I H****— exploring wire or cord* ***I****— hood* ***a****— percussion piston* ***b****— 3 sensitive primers* ***c****— gunpowder* ***d****— tight piston* ***e****— 6 lb ball & its cup*

fashioned as to resist being crushed, advantages possessed by a tin cylinder with conical ends & small diameter, though for small distances below the surface the form in plate 6 fig. 1 will answer all purposes, provided the tin be of the proper number or thickness.

The capacity of the tin torpedo should be about 8 gallons and its diameter lessened and length augmented for all depths, where it is likely to be submerged more than 10 feet.

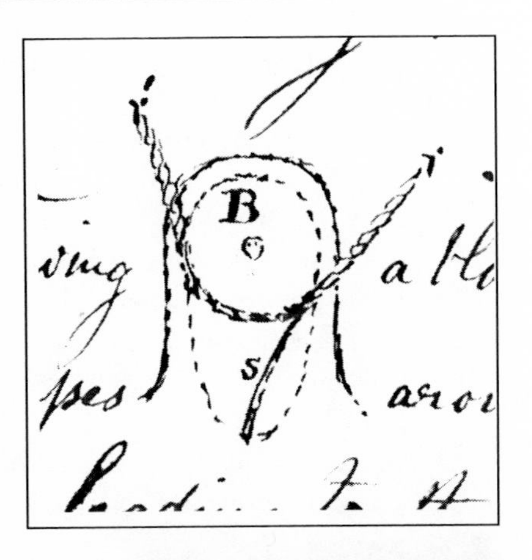

*Two anchors each having a block **B** for the rope (**r**), and a spring (**s**) to press against the rope as it passes around the block, can be connected with the staff of the torpedo higher (**r'**) leading to the staff and (**r**) to a boat.

[23] Demijohn Torpedo

This is quickly and easily made from any demijohn of sufficient capacity of 6 gallons or 8. It will hold an ample quantity of gunpowder from 50 to 75 lbs. but less will suffice, leaving air space for buoyancy.

To construct the torpedo, leave the basket upon the glass, and pour in about 40 lbs. of powder, if the demijohn be of the larger size.

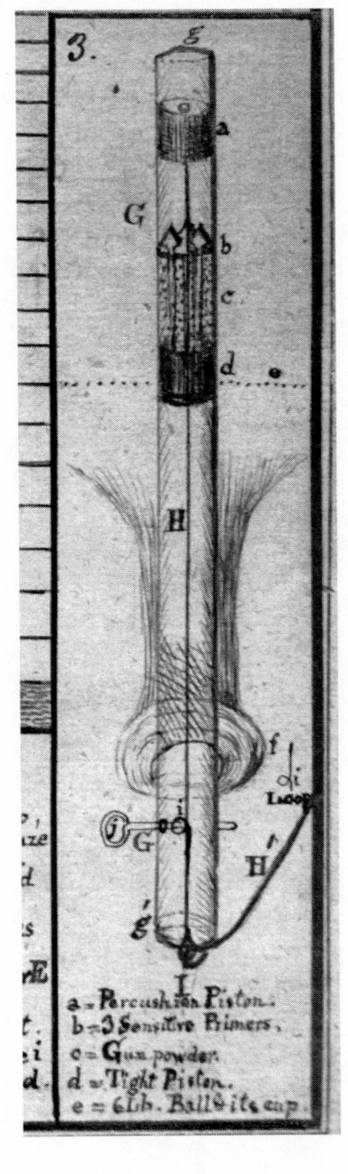

Then have a tin tube so made as to reach near the bottom of the glass with a button (**g**) (plate 6, fig. 3). This tube is to be air tight as well as water tight and is to be a stopper to the vessel made with gutta percha.

Into this tube a wire (**H**) with a block piston (**a**) fastened on its end is to be passed working loose somewhat in the tube, and three sensitive primers made up as indicated, their shanks passing through a circle of wood or metal at (**b**) their lower ends resting on another circle (**d**) and paper wrapped around making a cylinder or cartridge with gunpowder filling the space with the shanks of the primers between the two circles, and a small cylindrical tube in the middle for the wire **H** to pass through.

The cartridge is to be prepared before it is put [24] into the tin tube, and when ready, it is to be slipped on to the wire, and pushed home to its place at (**c**) fitting tightly.

The wire terminates in a loop (**g'**).

Now after warming the gutta percha on the tube (**G**), pass it into the mouth of the demijohn, the end (**g**) into the gunpowder, and cement it water tight in the neck by means of gutta percha.

Lash the anchoring rope to the neck of the torpedo, and its other end to its mushroom anchor at the length desired for the depth of water. The cup (**e**) is to be made fast to the bottom of the basket in position and the torpedo carried where it is to be planted (fig. 2)

With a cord doubled by passing through the ring bolt of the mushroom anchor (**E**) lower that so as to buoy the torpedo in proper position in the water, then hook on the wire (**H'**) into the loop at (**g**) with the 6 lb. ball attached to the other end and after placing this into its cup by means of the double cord lower the anchor carefully to the bottom of the water.

Finally let go one end of the cord, and pull it in by the other & row away from the spot. Should a vessel in passing come in contact with the torpedo, it will tumble over the [25] 6 lb. weight, and that will extend the wire (**H'**) which will pull upon **H**, and the piston (**a**) will be brought down upon the primer and explode them, and these the gunpowder, which will burst the tube & fire the torpedo.

The inversion of the demijohn & tin tube prevents the possibility of a leak, and it is desirable that the powder be kept out of the neck for the purpose of keeping it dry; where material is not at hand to make it sure, a little cotton will do it. The water then can only rise up in the neck wetting part of the cotton below, as far as the pressure of the atmosphere within will permit.

A demijohn makes a very effective torpedo as we learned from experience, and one charged with guncotton very superior (with the arrangement by me here presented).

Observations

An enemy's war vessel [*Cairo*] was destroyed in the Yazoo River by a demijohn torpedo. Capt. [Zere] McDaniel operator.

In an emergency torpedoes might be made of demijohns quickly and effectively (equal to those which destroyed 6 out of 12 vessels[14] which attempted to take Fort Branch up the Roanoke River in N.C.) With their baskets on, they are well protected, and the safety of this arrangement can be assured by a style or pin made of wire inserted through the tin tube & a loop (**i**) in the wire (**H**) to be extracted by a cord at the last moment to (**j**).

[26]

Aggressive Torpedoes — to go & anchor where desired (plate 7, fig. 1).

A torpedo made after the plan noted in plate 1 fig. 1—with its anchor rope buoyed up by cork has its mushroom anchor suspended from an empty barrel by a cord tied to it, and passing through a staple driven into the bottom of the barrel. (This barrel is to be water tight, a whisky or vinegar barrel or wine hogshead will answer and is to be of such capacity with the weight of the anchor as to float ⅓ or a ¼ above the surface of the water.)

The cord after passing through the staple is to pass up over the chime of the barrel and its side, then over the upper chime and through the head of the barrel to the interior where it is made fast to cotton string made into quick match with wet gunpowder and then thoroughly dried.

This quick match is fastened to the bottom of the barrel by means of another cord which is fastened to the one prong of the staple (**e**) bent within, or passed through a hole in the lower head of the barrel knotted without—and this hole tightly plugged by a peg driven in it, or fastened to a nail driven in. As many strands of the quick match as necessary will connect the upper cord with the lower, so [27] to be strong enough to sustain the anchor through the head, over the chime & through the staple below.

A well timed slow match (**z**) is now connected with the quick match. One made of cotton twine such as merchants use for tying up goods, doubled until 3/10 of an inch in diameter, will answer the purpose very well without further preparation, except a slight twist.

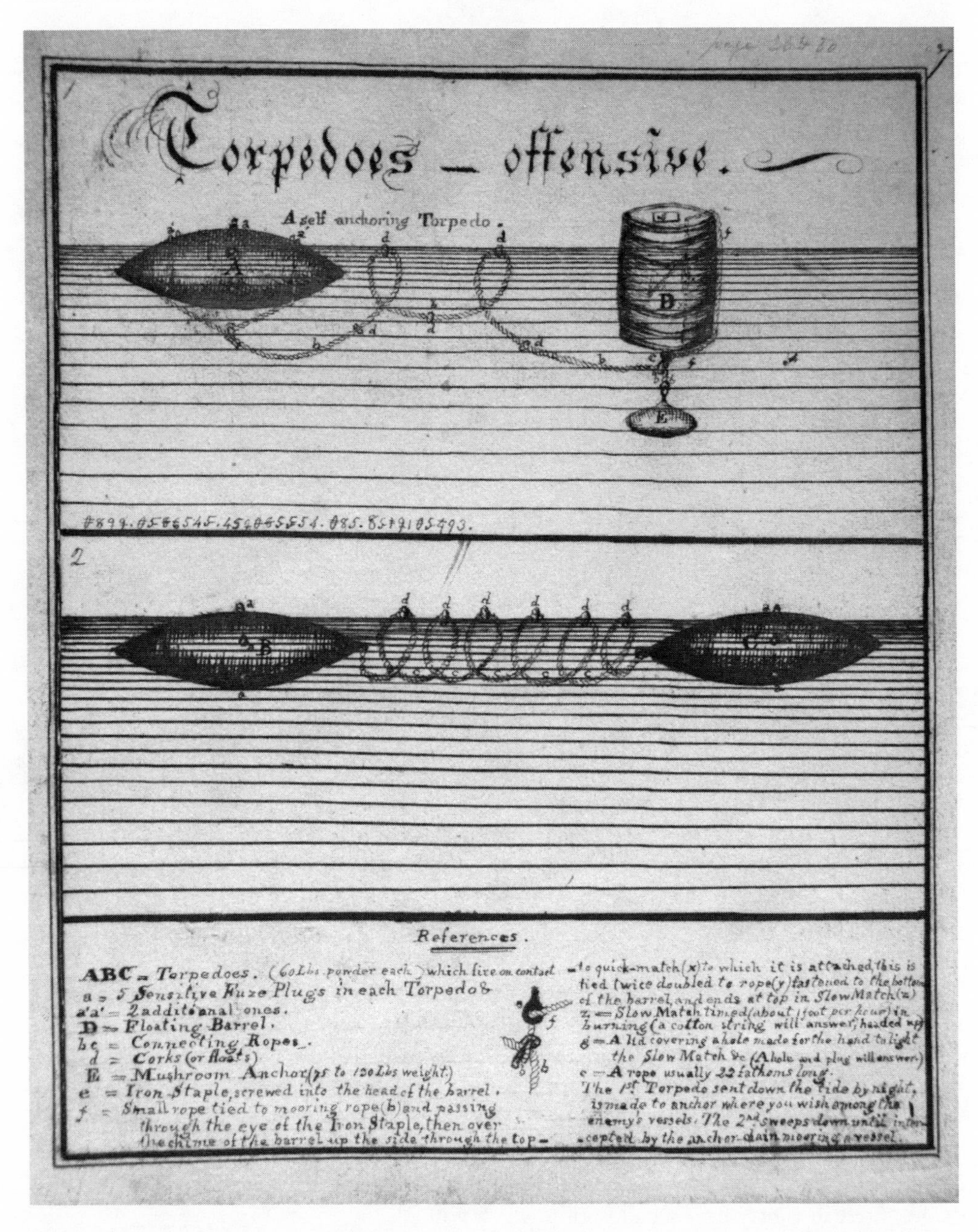

*Plate 7: References: **ABC**— torpedoes (60 lbs powder each) which fire on contact **a**— 5 sensitive fuse plugs in each torpedo & **a'a'**—two additional ones **D**—floating barrel **bc**— connecting ropes **d**— corks (or floats) **E**— mushroom anchor (75 to 130 lbs. weight) **e**— iron staple, screwed into the head of the barrel **f**— small rope tied to mooring rope (**b**) and passing through the eye of the iron staple, then over the chime of the barrel up the side through to the quick-match (**x**) to which it is attached, this is tied twice doubled to rope (**y**) fastened to the bottom of the barrel, and ends at top in slowmatch (**z**)*

***Z**— slow match, timed (about 1 foot per hour) in burning (a cotton string will answer, headed up) **g**— a lid covering a hole made for the hand to light the slow match &c (a hole and plug will answer) **c**— a rope usually 22 fathoms long*

The first torpedo sent down the tide by night, is made to anchor where you wish among the enemy's vessels. The 2nd sweeps down until intercepted by the anchor chain mooring a vessel.

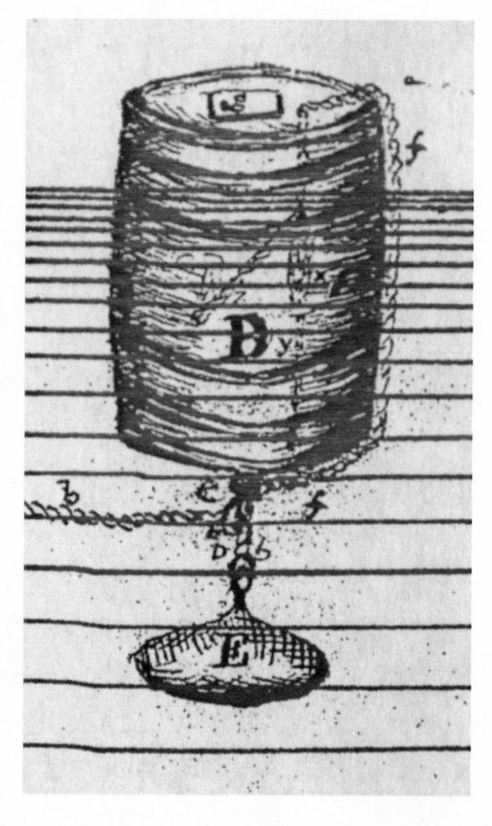

The slow match used for this purpose in Charleston harbor, burnt one foot an hour, and one foot is amply sufficient usually, which in a current 3 miles per hour would last that distance. After the matches have been arranged, the hole made in the head to insert the hand for this is to be closed with a sliding door which turns around on the head by means of the bolt (**g**) made water tight by a leather facing (a hole bored through the head for the end of the match to protrude so as to be set on fire, then pushed back & closed with a plug, will answer). It has been found that a barrel of 35 gallons contains enough oxygen gas in the atmosphere within, to keep the burning match alive for the above distance without ventilation, & for hours after. Even after the place has been swept for torpedoes by the enemy at night, a number of these can be turned adrift to be carried down by the return tide, and the slow match in length so arranged as to anchor **[28]** them among the enemy's fleet, or blow some of them up in their route. (The match must be a small cotton thing if to burn many hours.) The cut [*sic*] on (plate 7, fig. 1) exhibits one of this kind floating with its barrel anchor support and when it reaches the locality intended the slow match having burnt up to the quick match fires that which letting the cord loose holding the mushroom anchor, that immediately sinks to the bottom and the empty barrel goes on its way to sea. This barrel can easily be made to be blown to pieces by having a musket cartridge or two connected with the quick match if so required.

Of course the night is the time to send down such torpedoes, when there is no moon to betray them, nor enemy with lights on the look out. Both the torpedo and its anchor, with its rope coiled up, can be fastened at the bottom of barrel (**D**) by the cord which passes through the staple entirely out of sight under water, and thus can be sent to be anchored where desired.

To prevent taking up this barrel with the torpedo attached with impunity, a cap fastened to the upper head of the barrel thus may have a ball within which will fall out if decanted from the perpendicular, and so arranged with a string at that movement to pull down a strap upon a sensitive primer by which **[29]** fire will be communicated to powder to blow up the barrel and thereby anchor the torpedo or destroy the whole with the adventurers.[15]

A similar device can be attached to any torpedo at the lower part under water, to pull down a lever or strap upon the copper cap of one of the sensitive fuse plugs above, and explode it, should attempts be made to remove it from the water. This self anchoring torpedo was found to be a very efficient machine and by means of it, all attempts of the enemy to clear the stream of torpedoes can be frustrated and their vessels driven from their anchoring ground.

Besides this advantage of self anchoring — with its rope and barrel it is liable in passing through an enemy's fleet to encounter a ship and destroy her.

Observations

But one of these was sent down in Charleston harbor and that or the submarine boat (see [manuscript] page b) blew up the *Housatonic* war ship. Previously a barrel floating was dropped from Fort Sumter and its progress watched with a spy glass, and the time noted when it reached the enemy's fleet during the day and the slow match was lengthened accordingly and the torpedo sent adrift from that fort among the enemy's vessels and about the time required the war vessel was blown up. This torpedo contained 60 pounds of powder.

[30]

Offensive Self Acting Torpedoes (plate 7, fig. 2)

These consist of two torpedoes joined together as indicated with a rope buoyed upon the surface with corks and about 132 feet long, to go down or float by the tide. It becomes necessary to use these torpedoes in dark nights, to stretch the connecting rope in front of the enemy's shipping by means of two boats, and then to suffer them to go down by the tide. The arrangement of the sensitive fuse plugs about or rather around the middle of each enables them, when they strike a vessel, to explode readily, and the wide sweep of the rope between them rather ensures success by being intercepted by the anchor cables. It is believed that 50 of these so fixed would clear an anchoring station in a single night, and all vessels except ironclad must fall before their destructive effects. Two demijohns weighted about the necks joined with a cord will answer as in plate 6 fig. **F**.

The diagram sufficiently explains itself, with a remark that these torpedoes should just float and no more, having their specific gravity nearly that of water. So the charge of gunpowder should nearly fill them or otherwise they must be weighted by leaden bullets or other matter as iron spikes, sand, &*c*. which can be passed into the holes left for loading, & these might be enlarged some for that purpose if necessary.

[31] These torpedoes would strike an ironclad ram harmlessly, because as they are floating so near the surface of the water, they would necessarily strike against the inclined plane of thick iron covering, where their explosion could only jar the boat without destroying her. All other vessels of any description whatever must yield before this torpedo & its arrangement, which is simple, easily understood, are carried out.

Where these are ironclad vessels the arrangement on plate 8 becomes necessary.

Observations

I proposed to open Charleston harbor by these, to Chief Engineer [Col. Jeremy F.] Gilmer, but he dissented, to my surprise, asserting that they might come back into the harbor & blow up our own vessels or steamers. In vain I assured him, if sent at night, they could be seen when returning, as they must, by day, & the casualty prevented. Moreover, two rivers disemboguing (the Ashley & Cooper) would prevent, yet it all was of no avail, but determined to carry my point so self-evident, I went early to consult with Genl. [P.G.T.] Beauregard, but met Genl. Gilmer coming out of his office, where he had been to talk him out of it, which the Genl. was weak enough to permit. Query: Was Genl. Gilmer disloyal to our cause, or as I ranked him in the U.S. Army, did his opposition arise as with some others, from a mean jealousy of my success?

[32] The Torpedoes could be let go in half mile of the enemy's fleet — tide ebbing (plate 8, fig. 1)

These torpedoes are easily made, and are very simple. Two empty barrels for floats and joined together by a cord about 132 feet long (**bbb**) with corks or floats marked (**c**) which keep the rope on the surface of the water.

These barrels are to be nearly submerged by two torpedoes (**dd**) and sand put in each barrel, so as to bring them down nearly to the surface, when floating.

The suspending cords (**ee**) are to be 10 feet long each, and are to be made fast to the middle of the barrels as indicated in the diagram.

The torpedoes (**dd**) are to be loaded within, as to be but slightly or greater specific gravity than water, in fact just so as to sink in position. The object of this is that they may assume an oblique position under the influence of the current under the vessel's bottom whenever the barrel float becomes stationary by obstruction therein, at the side of the vessel.

Some 5 or 6 sensitive fuse plugs are put around the middle of each torpedo.

At night these torpedoes are to be taken in two boats (one in each) to the place where it is desirable they should be let go, and where they will be carried down by the current to an ironclad or [33] the fleet of the enemy & strike her in the bottom.

After stretching the connecting rope across the current, at a signal given, they let go the barrels (**ff**) and the men row the boats away. These barrels will float down with the current supporting their torpedoes below until they pass the anchor chain or cable mooring the ram when the joining rope will be intercepted thereby and each barrel tend to assume the position (**f**) as indicated in (fig. 2) i.e., against the iron shield of the boat.

As soon as the floating barrels are thus halted, or either one of them, the current carries the torpedo below it to the position (**g**), against the bottom of the ram below its iron covering where it immediately explodes, and sinks her by driving in or tearing off the bottom.

By fixing 3 or more torpedoes in the above fashion, after destroying one boat they are still effective for others, and a long line thus arranged, it would be almost impossible for it to fail under the direction of any ordinary intellect which could observe the way a current was tending.

Where there are but two, and one of these have blown up a boat, the other like a wandering knight goes about the water seeking adventure, and if intercepted [34] while going to and fro roaming the sea highway its torpedo, like the arrow in Achilles' heel, deals destruction to its opponent.

It is said that ironclads conquer the world but these torpedoes conquer the ironclads, and thereby discomfit all efforts of an invading foe to capture and destroy. They hurl back

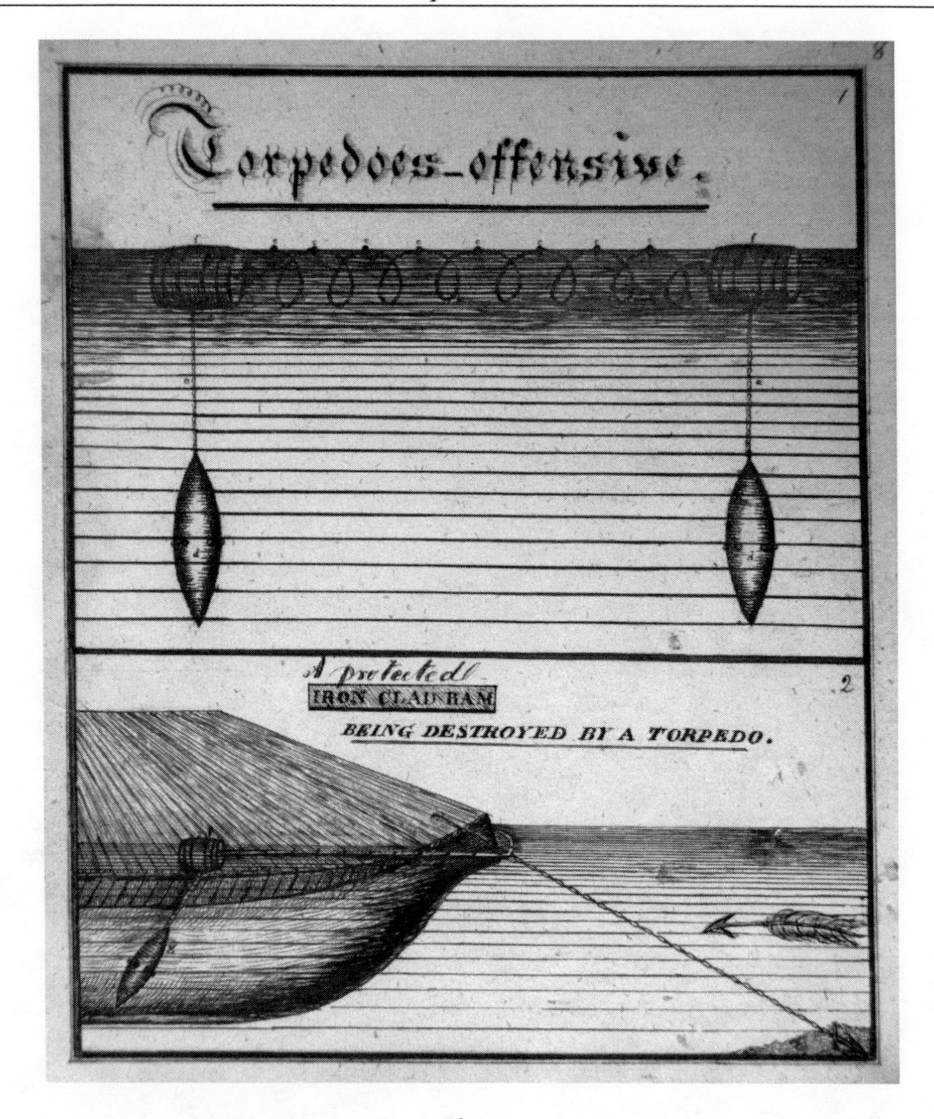

Plate 8

defiance in his teeth, and some 50 pairs of these at the entrance of each harbor always kept on hand ready for service at the nearest fort, with some ½ dozen other, particularly the submarine mortar battery, secures forever that harbor from invasion.

The cost of these torpedoes is but little, and nothing to be compared to that of interior forts which all, by the use of these, can be dismantled (in plate 8, fig. 2) is represented a torpedo striking under the influences of the current the bottom of ironclad ram, the very moment preceding its destruction. The barrels have been carried down by the tide with their torpedoes, and the first one which strikes sinks her leaving its fellow pass by breaking to pieces the barrel with the boat.

No fleet of vessels could enter a harbor where they had these to contend with by night & delay by day to take them out of the way under the guns of a fort might be fatal, and they could not proceed without they might be made to explode in being lifted out of the water.

[35]

Three Kinds of Torpedoes
Plate 9

The first of these we will describe is that known at Singers torpedo [designed by E.C. Singer and Dr. J[ohn] R. Fretwell (fig. 1 & 2). It is made of tin and of the shape shown in the diagram much like a street lamp in appearance at a distance. The thicker the tin the

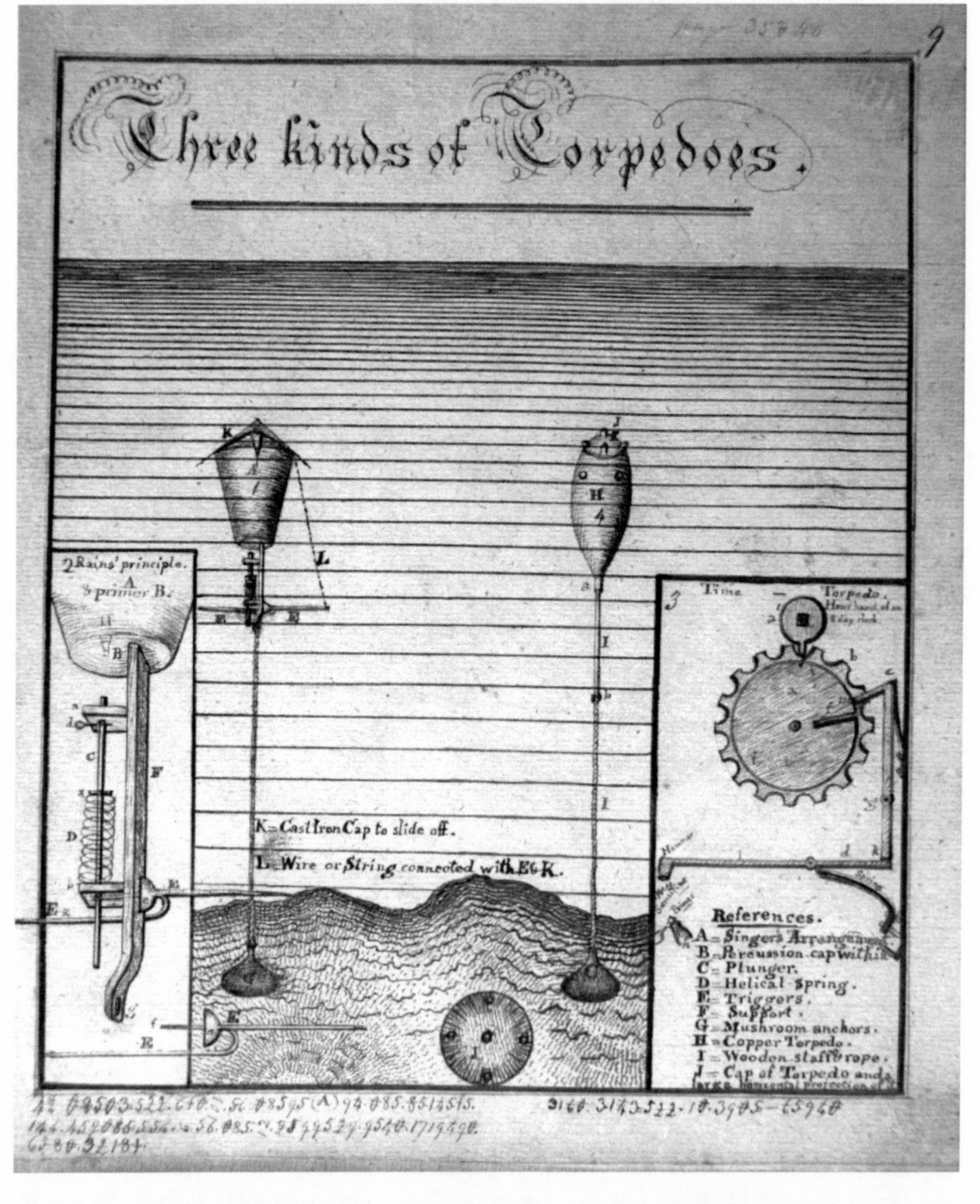

Plate 9: References: ***A****— Singer's arrangement* ***B****—percussion cap within* ***C****—plunger* ***D****— helical spring* ***E****— triggers* ***F****— support* ***G****— mushroom anchors* ***H****— copper torpedo* ***I****— wooden staff & rope* ***J****— cap of torpedo and a large horizontal projection*

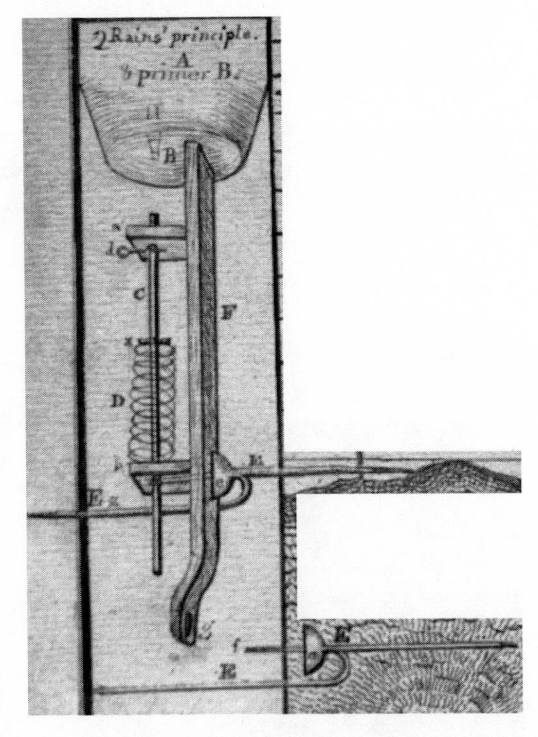

better to resist the pressure of water, which its shape is also calculated to do, being an inverted frustum of cone, topped by a cone, all of tin, the last cone being an obtuse angle one which is soldered to the inverted base of the frustum. (It may be strengthened here by a ring of metal if necessary.)

At the bottom there is extended on a brass projecting piece of the shape marked (**F**) in fig. 2. This has two projecting shelves like [*sic*] with a hole in each (see **a** & **b**) for the admission of the plunger. This plunger (**C**) has a cross pin (**x**) against which a strong helical brass spring (**D**) presses — the other end of the spring being against shelf (**b**).

A safety pin or wire (**d**) presses through the plunger as indicated which is extracted when the torpedo is planted.

The shape of the trigger (**E, E**) is separately shown below in fig. 2, but in position just above where it [36] passes through a hole in the brass stand (**F**) and also the shelf (**b**) into a hole made in the plunger by which, when set, the plunger is held in position. This torpedo is anchored with a mushroom anchor like the others, the upper end of the anchoring rope or cable being tied after passing through the hole (**g**).

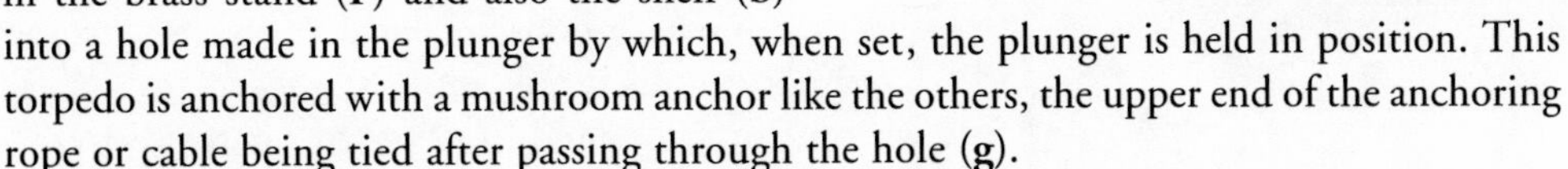

Supposing the torpedo planted as in fig. 1, a vessel passing over it pushes it down until its bottom comes in contact with the triggers (**E**) which acting as a lever round (**e**) as a fulcrum the point of it is drawn out for the plunger which immediately descends forcibly & impinges upon the disk (**B**) which is the thin bottom of the torpedo behind which inside the torpedo is a common percussion cap on its nipple fastened within and thereby explodes it.

The primer known as Rains' sensitive primer is a good substitute for the cap & nipple & is more certain.

A further improvement in this torpedo was made by the cup and ball as seen in plate 6, fig. 2 marked (**e**) — the ball being attached to the lever arms of the trigger, also an iron wheel to be pushed off. This has been found to be a very effective device in fresh water — but entirely ineffective in salt water from barnacles, and small oysters, soon getting [37] upon and fastening themselves upon the plunger and spring which prevent their action.

The torpedo is well coated with paint to prevent the action of the water (galvanic) upon the tin. Salt water soon corrodes it notwithstanding and renders it useless.

Remarks

This torpedo was used in Roanoke River having been planted there on an occasion, where they proved their efficiency by destroying 6 out of 12 vessels sent against Fort Branch [see manuscript page 25]. It is made upon my principle of impact upon a thin metallic plate excluding the water & coming in contact with a cap or prepared primer. My sensitive primers were used by Dr. Fretwell in preference to all others. The complexity of the outer works are objectionable — a fish &c. can set them off (see remarks elsewhere) and a strong current lean them down so as to drop the iron plate, when so arranged & thereby prematurely discharge them, though this latter may be remedied.

I sent Dr. Fretwell[16] down from Richmond to Roanoke River & he arrived in time, and according to my directions he succeeded in planting the torpedoes in the river above the enemy.

[38]

Time Torpedo (plate 9, fig. 3)

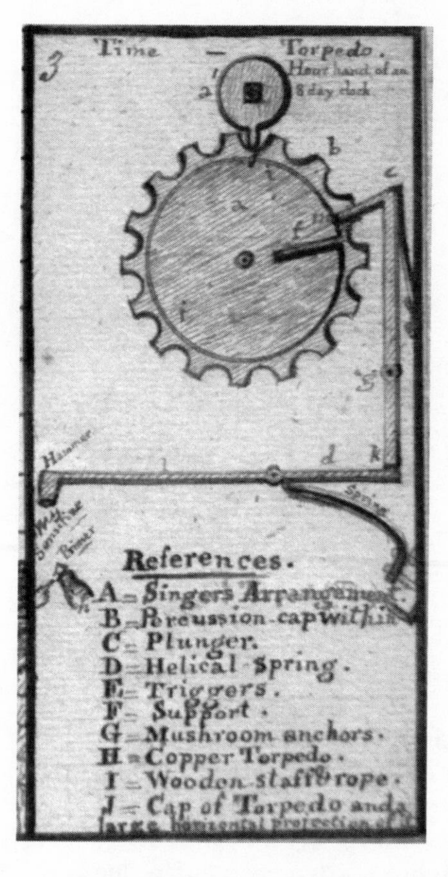

This torpedo can easily be comprehended by inspecting the figure. The wheels (**a, a'**) are a clock arrangement, the large wheel (**a**) having 16 teeth and the small wheel (**a'**) which latter turning on the 'shaft' of the hour hand goes around once in 12 hours or twice in a day and night. Each time the wheel (**a'**) turns around its tooth (**j**) passes between the teeth (**b**) of the large wheel and urges it forward one tooth so that this wheel turns once in 8 days.

The projections (**m**) on the lever (**c**) presses on the circumference of a circular disk attached to the large wheel until it falls into the vacancy or cut (**l**) when it turning round from left to right, sinistrosal, destrosal — as it turns around the center of motion (**g**) it detaches the end (**k**) by the action of a spring behind, and as soon as the end (**k**) slips beyond the end of lever (**l, a**) the spring near the fulcrum presses on that end and brings the hammer at the opposite end down upon the cap or sensitivity primer to explode it. By the fire of the primer the magazine with which it communicates is exploded. This arrangement can be easily simplified by using the sensitive primer in various ways. [39] It is not a new thing except the sensitive primer, though claimed both for originality and design. A wire protruding into the clock works, and extending to the outside of a box, magazine, bale of cotton, or goods of any description, is all that is required to start the clock, and the executer may be miles away when his infernal apparatus exhibits its effects. A marine clock that a twisted motion will set going or stopping will be better.

Observations

Capt. [John] Maxwell reported that he contrived to enter within the enemy's lines, at City Point on James River, below Richmond [August 9, 1864]; and finding a boat about leaving the wharf or dock for a vessel loaded with ammunition, he sent a time-box torpedo on board as something for the captain, who happened not to be there. In some 2 hours or less it exploded, set fire to the ammunition, & the boat blew up & fired her consort along side, which did likewise blew up & also the warehouse on shore. The explosion is represented [as] awful indeed, and thereby Genl. [Ulysses S] Grant's army was deprived of ammunition to fire upon Petersburg, Va., for one week. Then was the time for Genl. [Robert E.] Lee to have attacked but he did not know it perhaps. I have learned since that Genl. Grant & family came near being destroyed.[17]

[40]

Copper Torpedo

(Plate 9, fig. 4) [originally manuscript frontispiece]

Among the devices finally we come to this, probably the best of torpedoes, being also one of the most simple and easily constructed. It is made of copper sheets elliptical in shape

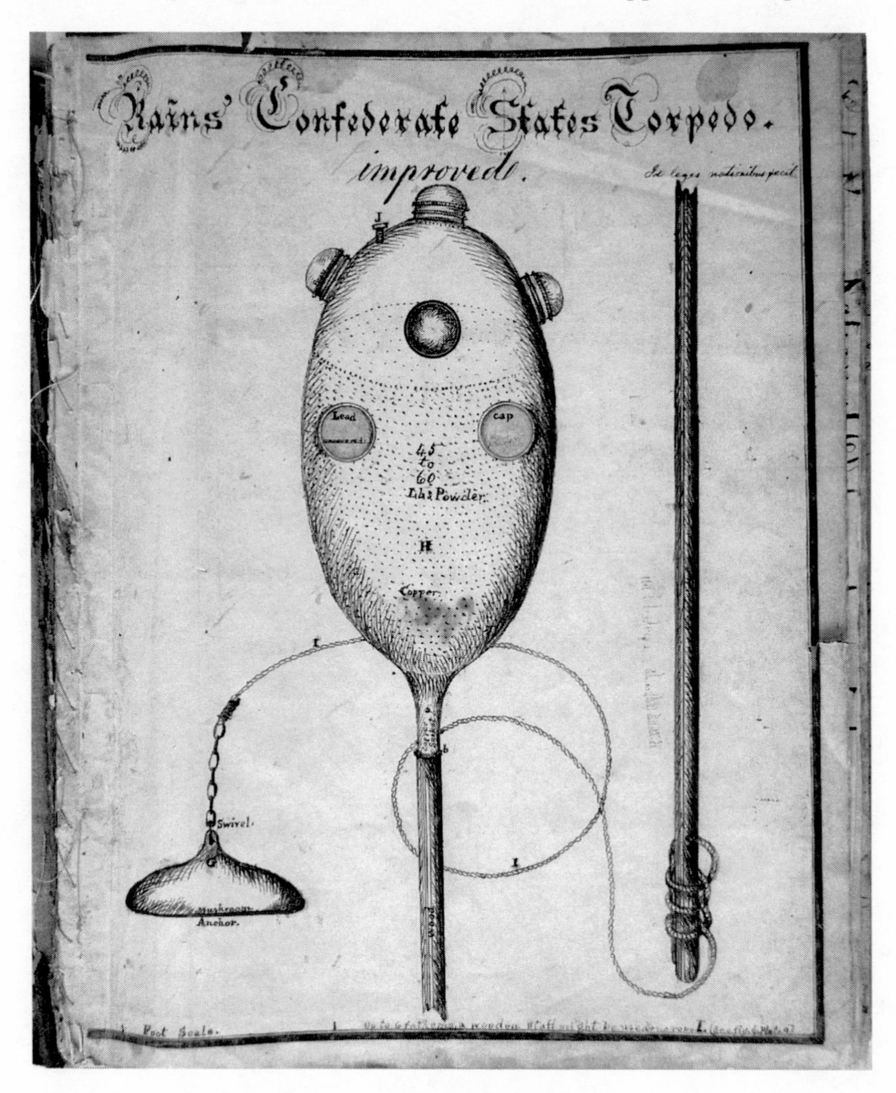

Frontispiece

and of capacity sufficient for from 40 to 50 lbs. of gunpowder and an air space above to float it. The lower end is drawn out somewhat after the figure where a socket is made (**a**) for the staff (**I**) and the cable rope (**I**) is made of the necessary length so as to bring the torpedo (**H**) within about 3 feet of the surface of the water at low tide when planted.

Where the water is but 25 feet, and under, a pole can with advantage be substituted for the cable (**I**) fitting into sockets at (**b**) and (**G**) and in this case, with two ropes fastened at the socket (**a**) one extending up stream and the other down, at an angle, with an anchor at the end of each, this machine becomes almost perfect in its kind.

There are at least 5 sensitive fuse plugs soldered into the oval top. The copper or leaden cap of each so thin as readily to yield to a hard body placed on its top while displacing the torpedo or pushing it aside. On account of the objections found to exist [**41**] from barnacles and small oysters in salt water a preference is given to the sensitive fuse plug in figure 1, plate 2 — with its hexagonal base (hexaedron) to screw it into the copper top of the torpedo or[18] a brass female screw soldered in.

The copper cap **G**, or better one of sheet lead, can be made so these are connected as to yield to any force which would ordinarily be applied, or would set off any other means, and thus becomes simplicity itself with the three sensitive primers beneath it. Of course no sea barnacles fasten upon the copper, or if such were the case no obstruction of this kind would prevent efficiency (see front piece). The advantages of this torpedo over others are many, they are so easily constructed (I have had 34 made out of one copper still which had been employed for making spirits of turpentine & rosin) and if tight when properly tested for leak they will last until taken up or exploded, till doomsday.

Observations

58 vessels are said to have been destroyed by torpedoes including several ironclads.

[42]

Observations

I shall treat next of electric torpedoes or such as we exploded by electricity. I am <u>not in favor</u> of these kinds of torpedoes except in certain localities. The grand error has been committed by nations to depend upon setting off their torpedoes by electricity, at the proper time to be determined by persons on shore.

Besides the depth of water rendering them ineffective, the agitation at the time or from some other cause, it has been found quite impossible to fire them at all or exactly at the opportune moment and for these & other reasons I have determined it best to require actual contact of the enemy's vessel to explode them, and experience has proved my surmise correct. Hence the secret of my success.

Torpedoes to be exploded by electricity for vessels, will fail to do execution or to be set off by strings or wires, as at Fort Caswell, N.C., where the enemy's shot cut them, also at other places in Virginia, where the operator ran away on the appearance of the foe and left them.

[43]

Galvano — Electric Torpedo
Plate 10

The submarine cable for the conduction of electricity to this torpedo is essentially composed of an outer coating of iron wire (**D**) wound around as indicated in the diagram (fig. 1) and an inner tube of gutta percha (**G**) enclosing seven small copper wires in a twist. (This is evidently part of the old [Trans-]Atlantic Cable which was brought to Charleston and used for this purpose.)

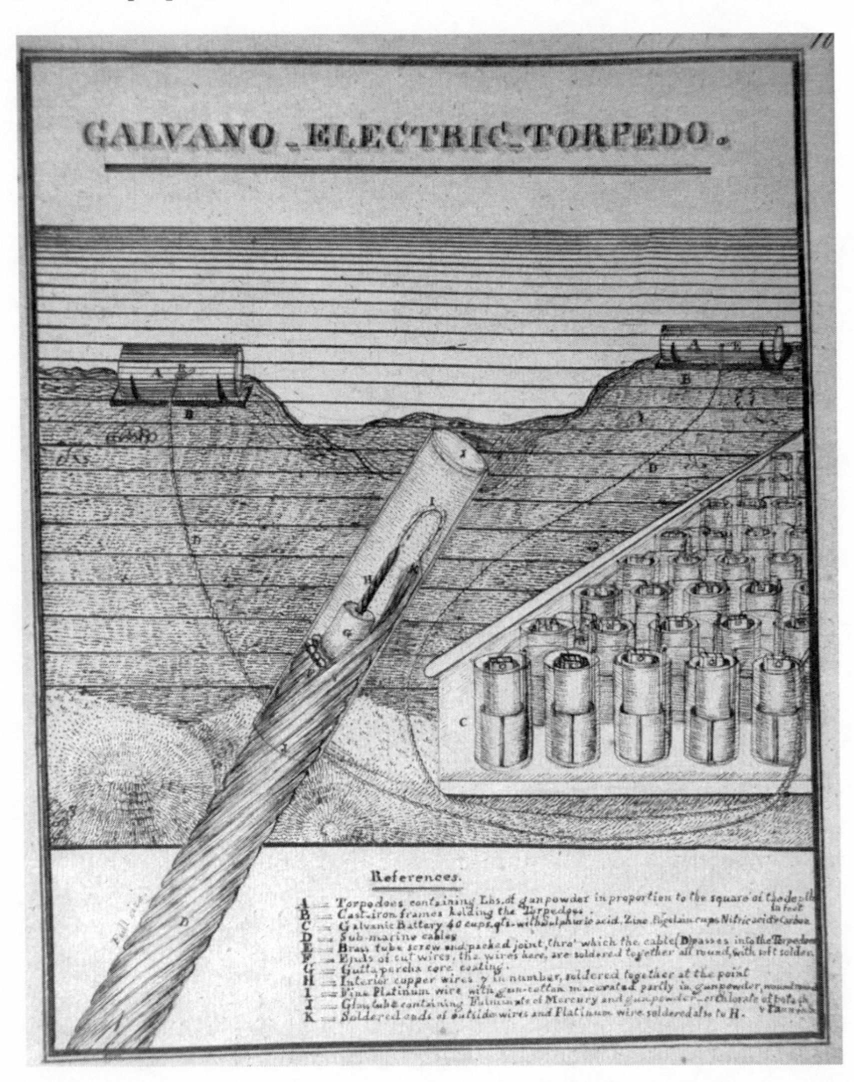

*Plate 10: References: **A**— torpedoes containing lbs. of gunpowder in proportion to the square of the depth in feet **B**— cast iron frames holding torpedoes **C**— galvanic battery 40 cups qts. with sulphuric acid. Zinc porcelain cups nitric acid carbon **D**— submarine cables **E**— brass tube screw and packed joint, through which the cable (**D**) passes into the torpedo **F**— ends of cut wires the wires here are soldered together around with soft solder **G**— gutta percha core coating **H**— interior copper wires 7 in number, soldered together at the point **I**—fine platinum wire with gun-cotton macerated partly in gunpowder, wound around **J**— glass tube containing fulminate of mercury and gunpowder — or chloride of potash & [illegible] **K**— soldered ends of outside wires and platinum wire soldered also to **H***

To prepare it for service, part of the covering wires are filed off at (F) and some 3 or 4 of the remainder soldered together with soft solder and brought to a point at (K) where is also soldered a fine platinum wire about an inch in length, the other end of which is soldered onto the termination of the copper wires. A little gun cotton macerated somewhat in meal powder is wound around this platinum wire, the object of which is to take fire at a very low temperature. The tube (J) of glass is filled with either of the following, *viz.*

1st	pure gunpowder fine grain or	
2	fulminate of mercury	10 [parts]
	gunpowder mealed	5
3	chlorate of potash	10
	sugar	5
	sulfur	2
4	chlorate of potash	10
	prusiate of potash	5
	gunpowder fine grain	5
5	chlorate of potash	10
	glycerine	4
	gunpowder fine grain	4

[44] After it has been passed over the end of the cable as shown, it is now to be stopped up with a cork at (J) and the iron wires around at (F) are to be soldered together by means of soft solder (pewter) sal ammonia & rosin and the glass tube fixed in position by gutta percha or better this, put on the wires warmed, & the glass also warmed passed over it.

This is now ready for the torpedo which is a small iron boiler used, or intended for steam purposes, but without the flues. This boiler should be large enough to contain 1,500 or 2,000 lbs. of gunpowder for 44 feet depth, which is put into it after it is pitched within and without & made perfectly secure from leak. The end of the cable is passed into the boiler through a stuffing box where it is made tight with gutta percha and rope stuffing & the man hole closed.

The boiler is put upon a cast iron bed (B), where it is secured and then transported with the cable once reel[ed], to the spot where it is to be planted.

A way to transport the torpedo and locate it, if in the vicinity of the enemy, is to plant it on its bed beneath a float which can be made of empty barrels or hogsheads, and there so secure it by lashings that a single cut with an axe will liberate it to sink to the bottom. This plan was found to succeed well in Charleston harbor. [45] The boat or flat containing the reel and cable should in this case pay out [a] sufficiency of it for the torpedo to sink to the bottom in the direction it is to be laid, without a jar. The end of the cable is then brought ashore, and an electric battery prepared of 40 quart glasses with as many 'zincs,' porcelain cups, and carbons with affixtures as necessary. Sulphuric and nitric acids. Ten of these 'cups' in good order with the zincs amalgamated will suffice for rendering the platinum wire red hot for three miles (1½ going & returning) yet better there be too many than too few, and the number as above will be about right for that distance. An error was made in Charleston harbor that had as well be mentioned here to prevent like hereafter.

The ironclad [*New*] *Ironsides* stood or rather laid over torpedo for ½ an hour but the sparks of electricity nearly 6 inches long, from the Rhumkorff coil, were sent along the cable, it failed to fire the torpedo, though it required but the 30th part of an inch for passing between two wires imbedded in powder highly explosive within the torpedo. Suspecting the fact from the nature of the Rhumkorff coil which tends to change galvanic into static electricity, that had passed through the gutta percha coating rather than go the distance required at fartherest not over [46] 2 miles. I tested the fact and found cables through which electric sparks had passed from a Rhumkorff coil, entirely useless afterwards for submarine purposes. The outside wires of the cable were found adequately sufficient for the return current of electricity & the arrangement will work admirable — the iron wire without and copper wires within being connected with the zinc and carbon — or positive & negative electrodes.

This torpedo will readily close a narrow channel to an enemy and at the same time enable friend to navigate over it, yet in dark nights a calcium light or the ordinary parabolic mirror of a locomotive will answer in most cases, — or a Fresnel lens.

The objections to this kind of torpedo are — It is liable to leak under the great pressure of the superincumbent water. It is very apt to occur in a new boiler & until rust closes the seams or it is thus made tight by acids, or sal ammoniac, sulfur & fine iron filings rubbed into every possible crack.

An Indian rubber bag enclosing the charge of gunpowder perfectly water tight had better be used in addition to other precautions taken.

Another strong objection is found — in keeping the electric battery always ready — the zincs [47] amalgamated — the sulphuric acid (diluted with about 19 parts water for the zincs, but not poured in) & the nitric acid, which for a short time, say a day or two at a time, can be kept with the carbons in the porcelain cups, — but longer periods that substance, unglazed as they should be, & deteriorates its power. To keep the connections also perfect with the slips of copper and binding screws all together, is no little trouble and when the enemy's steamers are advancing, say at midnight, it is difficult to get all things ready for action in time.

Hare's Deflagrator would be a better instrument for ignition of the platinum wire within the torpedo than perhaps any other battery, for the above objections will equally apply to those of platinum foil and give for most of the kind now in vogue. So the diagram on plate 10 is delineated rather as the one used in the service of the Confederacy, than recommended for such purposes (where the depth of water is 30 feet it is not admissible). In Charleston, S.C., these torpedoes contained 1,500 and 2,000 lbs gunpowder each. There were 8 in number but two of them depended for ignition on static electricity as furnished by the Rhumkorff Coil, and but one of the plan represented which, however, was that used in Mobile harbor and in James River near Richmond, Va. [48] (The guncotton partly macerated in meal gunpowder has been since added as an improvement on this method of firing torpedoes or mines and that substance takes fire readily at comparatively lower temperatures as to be desired in such cases.)

Observation

Deserters soon informed the enemy of the above torpedoes & I did not object as it serves to frighten them off & I wanted the channel kept open for our own steamers in Charleston harbor.

Had the enemy come over them when fired it is possible they would not have been injured.

Continued on [manuscript] page 73 [*sic*]

To know the precise spot where torpedo is located, sight from two positions at angles with each other, or use a camera obscura from an elevated position, and mark on the ground glass the spot, which for this purpose might be indicated by a white barrel at the moment when the torpedo is sunk (which is afterwards removed of course) Should an enemy pass over the torpedo the ground glass will then show it, or a sheet of paper horizontal on a table, the picture being formed thereon from a lens & mirror above, so that the proper time for destruction will readily appear.
[49]

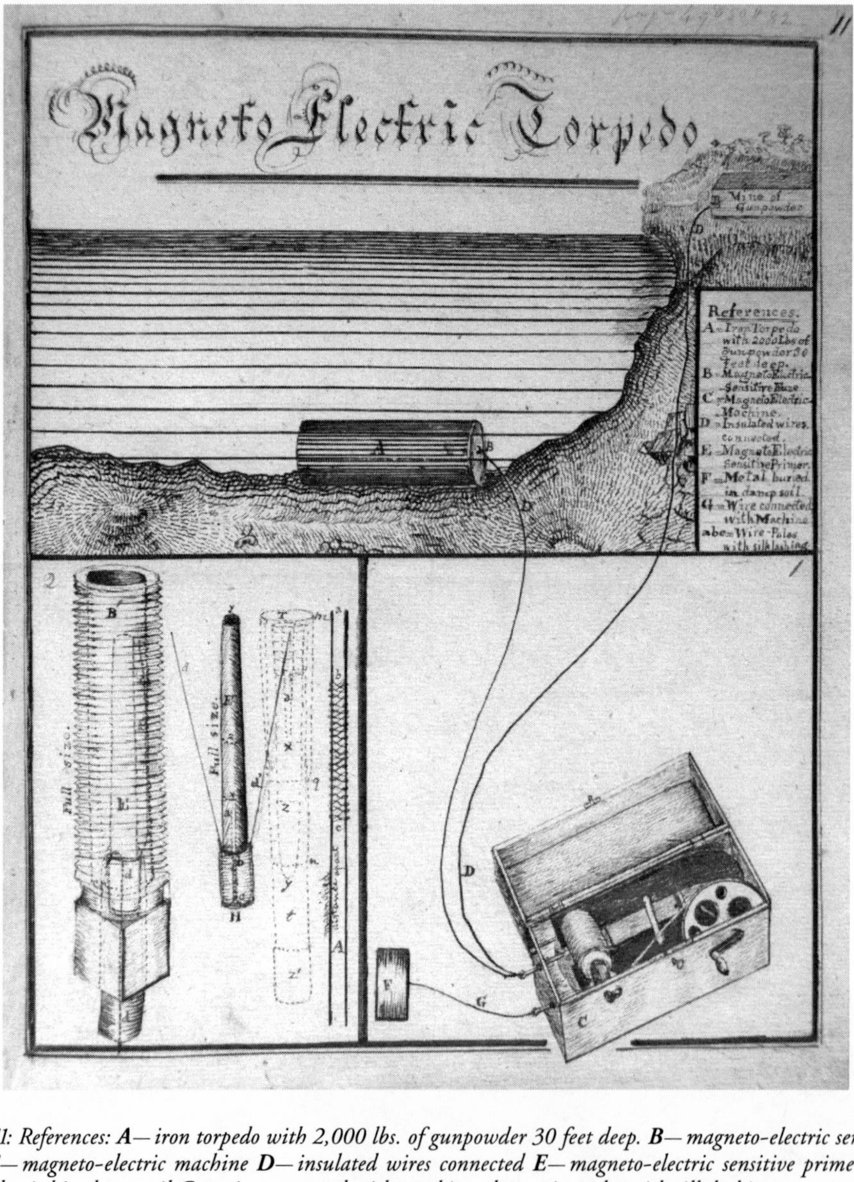

Plate 11: References: ***A****— iron torpedo with 2,000 lbs. of gunpowder 30 feet deep.* ***B****— magneto-electric sensitive fuse* ***C****— magneto-electric machine* ***D****— insulated wires connected* ***E****— magneto-electric sensitive primer* ***F****— metal buried in damp soil* ***G****— wire connected with machine abc— wire poles with silk lashing*

Magneto Electric Torpedo (Plate 11)

Of all forms of torpedoes which explode by electricity this is considered the best of its kind. It consists essentially of a torpedo made of a small boiler as in the last instance — containing from 1,500 to 2,000 lbs. of gunpowder with a magneto electric sensitive fuse plug (**B'**) inserted by being screwed into the end as at (**B**). This torpedo, as others of this nature, is placed at the bottom of the river or harbor and has a gutta percha or otherwise insulated wire from there communicating with the fuse plug above mentioned, and the other end on shore with a magneto electric machine (**C**) the other pole of which by means of the naked wire (**G**) is fastened to a plate of metal (**F**) buried in moist soil or instead connected with water or gas pipes leading under the soil.

The object of this latter is to use the earth circuit for the return current of electricity. The metallic magneto electric sensitive fuse plug itself & torpedo of metal affording the necessary conductors. Of this kind of torpedo (not the best) the great merit of this machine beyond all others for the electric communications consists in the sensitiveness [**50**] of the primer to electricity induced by magnetism. The plate 11, fig. 2 is drawn a full size primer (**E**) with its two poles (**d, e**) projecting at an angle from the sides, made of copper wire $1/80$ of an inch in diameter. This primer is seen in position (**E**) in the brass fuse plug (**B'**) and while one of its poles lies against the metal side at (**m**) the other (**d**) is bent down in the manner indicated and, perfectly insulated with gutta percha, passes out of the plug & is united to the insulated wire (**D**) by being soldered, care being taken that no part of this wire is exposed uninsulated.

The brass fuse plug, after the primer is fastened within, is filled with gunpowder around it and above and a cork stopper closes the hole there — at (**B'**) when it is ready for being screwed in as stated. In fig. (**A**) is seen how silk thread unites the two poles together — magnified, for the poles approach $1/20$ of an inch in reality.

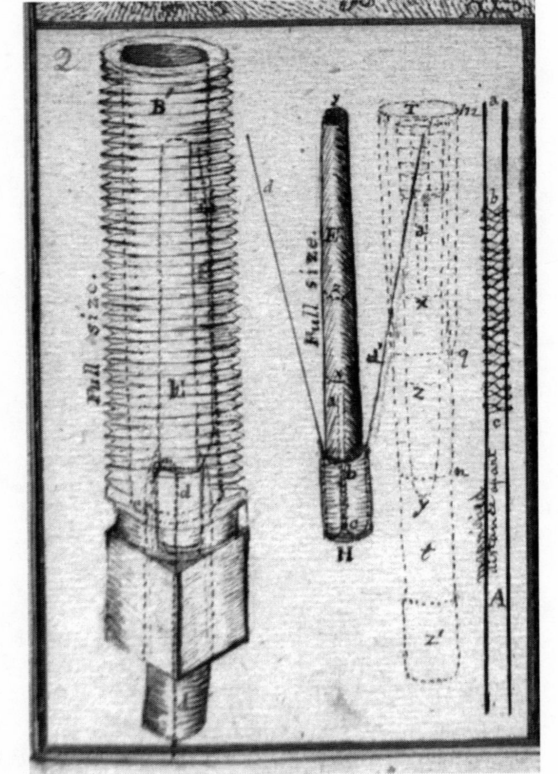

Plate 11, fig. 2

The advantage of this over other electric torpedoes consists in the facility with which the magneto electric machine is carried about and managed.

With the primer it may be made readily to five mines, or shells, under the covered way of a fort or at a landing which it is desirable should not be used by the enemy.

One small good strong magnet thus far has been found [**51**] sufficient with its two revolving helices & poles of soft iron to fire the magneto electric sensitive primer — yet for long distances, a magnet compounded of several might be used to advantage.

The composition and manner of making these primers is seen on [manuscript] page (82). (It is my invention also)—this is really a great discovery—as it reduces the firing of these torpedoes to great simplicity, and enables us to use these destructives to great advantage.

The magazines to [*sic*] all forts can now, by this means, be easily set on fire and the shouts of triumph of a successful foe end in disaster and death. No acids, no metals dissolving, nor jars of glass are required but a plain simple medico magneto electric machine in size but a few inches, is necessary which this invention renders effective in the hands of a commanding officer to blow up his magazines & destroy the successful enemy. This is the best means to fire by electricity, superior to all known methods, and its discovery by me ought to be a fortunate one.

A mixture of high conducting power & great susceptibility to ignition invented by Wm. Abel of Westwich[?] [illegible] consists of a combination of subphosphate of copper and sub[illegible] of potash exploded by a small electro-magnetic machine.

[52]

Observations
On the annexed subject of subterra shells

To checkmate any army however large, a *Corps d'Armee,* a large body of cavalry troops should be organized to hover in front of an advancing army of the enemy whose duties shall be exclusively to plant in the lines of their route the subterra shells. Doubtless a thousand men so employed should stop any army in the world—for no soldiers will march over land <u>supposed</u> to be mined & still less when a certainty crowns that fact. The cavalry force are to guard and protect the 1,000 shell men on this duty, and also to guard the shells themselves from friends when planted & secure the operators when laying down or taking up the mining shells.

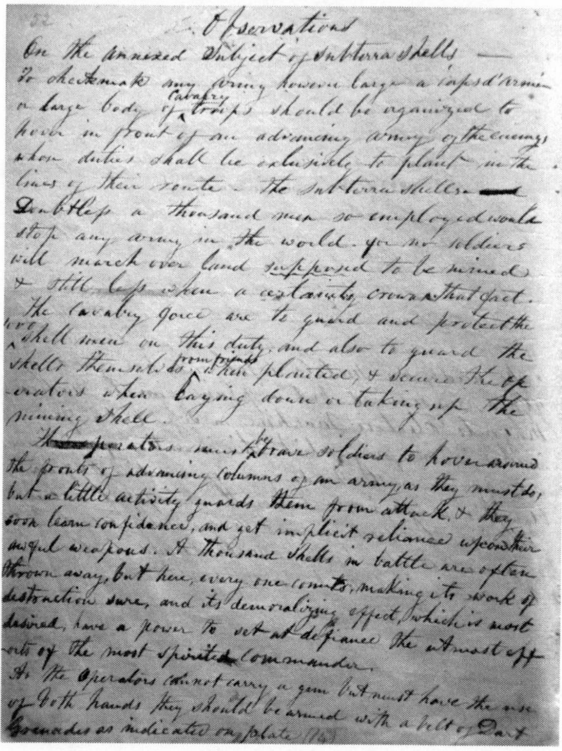
52
Observations
On the Annexed Subject of Subterra Shells —
To checkmate any army however large a Corps d'Armée
a large body of cavalry troops should be organized to
hover in front of an advancing army of the enemy,
whose duties shall be exclusively to plant in the
lines of their route the subterra shells —
Doubtless a thousand men so employed would
stop any army in the world — for no soldiers
will march over land supposed to be mined
& still less when a certainty crowns that fact.
The cavalry force are to guard and protect the
1000 shell men on this duty, and also to guard the
shells themselves from friends when planted, & secure the operators when laying down or taking up the
mining shell.
The operators must be brave soldiers to hover around
the fronts of advancing columns of an army, as they must do,
but a little activity guards them from attack, & they
soon learn confidence, and get implicit reliance upon their
useful weapons. A thousand shells in battle are often
thrown away, but here, every one counts, making its work of
destruction sure, and its demoralizing effect, which is most
desired, have a power to set at defiance the utmost efforts of the most spirited commander.
As the operators cannot carry a gun but must have the use
of both hands they should be armed with a belt of Dart
Grenades as indicated on plate (14)

Copy of Manuscript page 52

The operators must be brave soldiers to hover around the fronts of advancing columns of an army, as they must do, but a little activity guards them from attack, & they soon learn confidence, and get implicit reliance upon their useful weapons. A thousand shells in battle are often thrown away, but here, every one counts, making its work of destruction sure, and its demoralizing effect, which is most desired, have a power to set at defiance the utmost efforts of the most spirited commander.

As the operators cannot carry a gun but must have the use of both hands they should be armed with a belt of dart grenades as indicated on plate (14)[19]

[53]

My greatest discovery in War, which renders invasion impossible Sub-terra Shells or Land Torpedoes

Plate 12 [13]

This is regarded as the '*ne plus ultra*' of modern inventions for warlike purposes, for by it we can turn & checkmate mighty armies, and so demoralize soldiers as to paralyze all efforts for successful warfare.

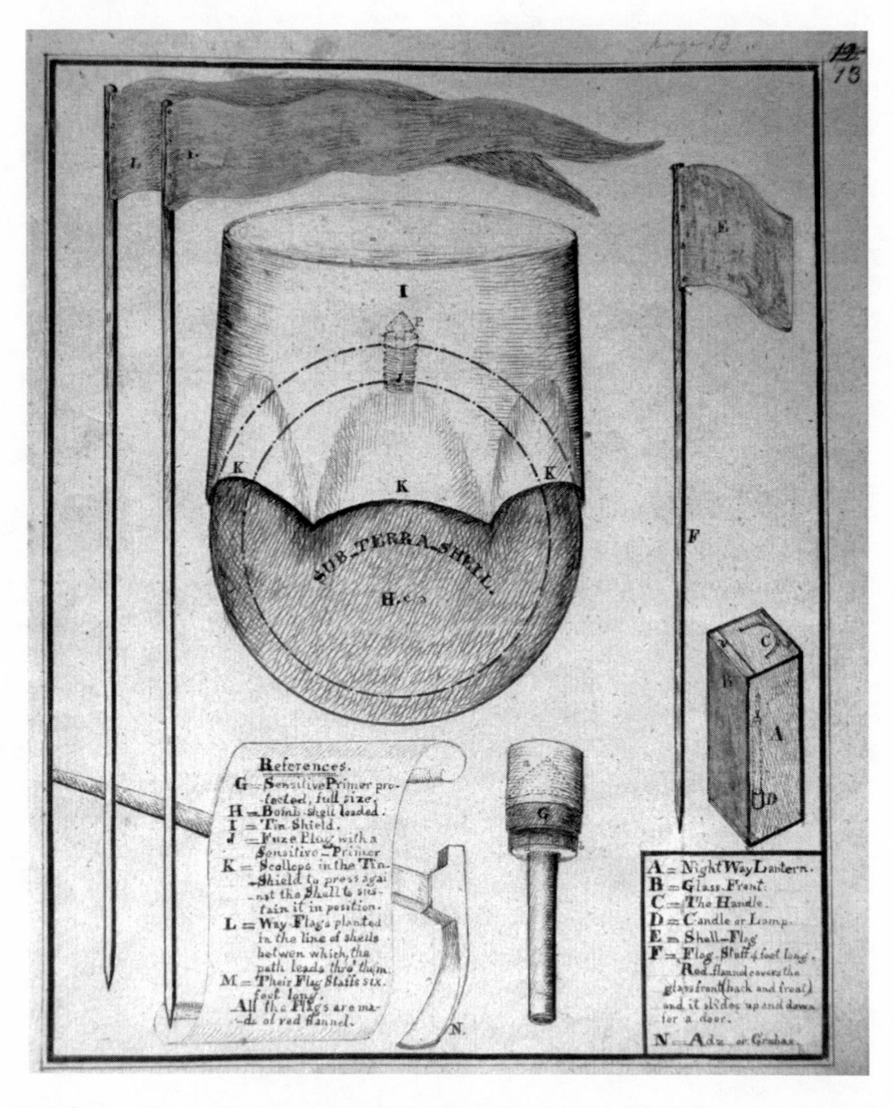

*Plate 13: References: **G**— sensitive primer protected, full size **H**— bomb shell loaded **I**— tin shield **J**—fuse plug with a sensitive primer **K**— scallops in the tin shield to press against the shell to sustain it in position **L**— way flags planted in the line of shells between which the path leads through them **M**— their flag staffs six feet long. All the flags are made of red flannel*

***A**— Nightway lantern **B**— glass front C— the handle **D**— candle or lamp **E**— shell flag **F**—flag staff 4 feet long red flannel covers the glass front (back and front) and slides up and down for a door **N**— adz or grubar*

By it, we can make our own forts impregnable and change the shouts of triumph of an invading foe to lamentations and wails of woe. The ethics of war is [*sic*] not determined by any fixed rule, to say the best of it, war is but murder legalized on a large scale — even massing and sending a hecatomb of human beings to eternity in a moment, is admissible, and this is maiming but on a smaller scale. Each new invention in the material of war has been assailed and denounced as barbarous and anti–Christian, yet each in its turn notwithstanding, has taken its position by the universal consent of nations according to its efficiency in human slaughter. A mighty revolution was affected by gunpowder, the bow and ballista with the tormentum give way to the musket and that now is being fast replaced by the rifle, the battering ram and catapult gave way to the cannon of smooth bore which in its turn is yielding to one rifled and the spherical to elongated shot. To prevent wars we must perfect the means of destruction [54] so as to render invasion impossible.

Did we have means to sweep a battlefield in a moment there soon would be no battlefield. If we cause the earth to open in a thousand places and destroy a host by fire and shells, soon no troops will be found to hazard a march over such a volcano. This we profess to have to accomplish, as has been shown by a proper use of subterra shells by means of which we can say to an enemy in thunder tones, "so far shalt thou go and no farther."

By these, the defense becomes all powerful, the offense nugatory, for no soldier will madly rush into the very jaws of death, when that is hidden and in mystery. Pure philanthropy led to this invention, for if by it we put a stop to all aggressive forays, hostilities become impossible and when universally known under the providence of God is as herein the dawn of that day when nations shall learn wars no more.

Gen. G. J. Rains

N.B. I went and examined the road on which McClellan's army was advancing upon Yorktown, Va., with a view of checking it but found citizens on the move, &*c.*, on mature considerations concluded not to use the subterra shells without due notification. After I had left Yorktown, for though I had been in command there my brigade was first to be moved, having the whole army under Gen. [Joseph E.] Johnston there [illegible] and some of the subterra shells improperly [illegible] they were falsely attributed to me. [55] The subterra shell is any bombshell, plate 12 [13] (**H**) planted under ground out of view plugged up with a wooden stopper (**J**) with a sensitive primer (**P**) inserted therein as indicated, so that a foot tread upon the top of it, explodes the primer and thereby the shell loaded with gunpowder.

This is usually buried so that the primer may be effective at all times, with a tin shield or water proof cloth (**I**) covering the upper part of the shell with scallops (**K**) around its edge fitting upon the surface thereof. The object of these scallops is that the inverted tin cup may, by the elasticity, fit down upon the shell, but when trodden upon they give way & allow the cup or shield, as it is called, to press down upon the primer over the shell. It is evident from the position of the shield that no water can reach the primer though it be planted under it, and as long as the shield remains entire the primer will be effective beneath the surface of the ground or water.

This shield before used is to be dipped in coal tar with or without rosin dissolved therein, and it is then to be rolled in the sand to take on the appearance of the earth as much as possible where it is to be planted. Only so much is removed of the earth as to admit the shell, and it is then to cover the top of the tin [56] shield about an inch or so, that no sign of the shell will appear — the rest of the earth must be carefully scattered or thrown away.

Some 2,363 mostly 24 pdr. shells were planted in front of the Confederate lines near Fort Harrison below Richmond, Va., and in the rear of each subterra shell, the flag (**E**) was raised 3 feet distance on a staff from 3 to 4 feet long—the flag being made of red flannel about 10 inches square nailed upon its staff. The object of this small flag is to keep enemy soldiers away from the shells, but they should be removed at night and replaced in the morning. They should also be removed on suspicion of the approach of the enemy. Their demoralizing is such as to need no concealment, for no enemy will charge over them at any time.

The two streamers on the long staves (**L, L**) are intended to be placed in the line of subterra shells and so planted as to mark the outlet between them. This passage way must be used by our own men—skirmishers *&c* and at night the Nightway lantern is hung upon the top of each to indicate the same. These lanterns must be extinguished at tattoo—or before if any indications are shown of an advance of the enemy. The lanterns are made of tin and are either for lamp or candle.
[57] The Nightway lantern is made from a sheet of tin by cutting after the pattern laid upon it, the two ends (**C & D**) having a vacancy (**V**) both at top and bottom, and the top (**C**) another next to the back. The piece (**C**) forming the top must be riveted on as well as soldered and the handle also must be riveted on.

To plant the subterra shells men must be selected of staid & sober habits, and drilled to it, to prevent the possibility of accident.

The bomb shells loaded—stopped with a wooden plug in the eye with a hole through it, into which is inserted a screw (or peg will answer, but not so well), is to be carried with the requisite number so prepared in a caisson, ambulance, or spring wagon to the place where intended to be used (or in bags on mules if necessary). A hole dug in the soil with a grubar [*sic*] (or a spade will answer) is to receive the shell, it is then filled in around the bottom leaving only the top exposed so far as is necessary for the tin shield. This should now be put on and pressed down upon the screw and a chalk mark made on the shell to show how far it can with safety be put on, after the sensitive primer is inserted—the scallops then must be reformed. The man to operate farther must kneel upon one knee—he is to unscrew & remove the screw—insert its primer, and put on the tin shield bottom up, then cover with earth or sand out of sight.
[58] These tin shields must be so far raised as judging from the chalk mark it can be safely done—he is then <u>with his hand</u> to put the dirt or earth around the shell and then covering the top of the latter a half inch or more. He now plants the shell flag (**E**) and retires. For night service or where there may possibly be any rough treatment the sensitive primer (**G**) plate 12 [13] is used.

This primer, the same as in (plate 1, fig. 2), has a cylindrical cover or top (**a**) made of tin, fitting tightly on the primer as shown and afterwards this is further secured by paper wet in mucilage wrapped around it, and an additional short cylinder of thick leather (**b**) pushed upon the shank by it. The object of this leather is to allow the tin cap cover, when trodden upon, to be forced down upon the cone of the primer to explode it.

Elongated shells and in fact all kinds of shells even most of those condemned under inspection will answer for subterra shells—holes or other imperfections being closed with pitch, gutta percha, or the like.

Where the shields are lacking any shell loaded can be used temporarily by burying the shell and sticking in a primer or two and covering up a board, shingle, or something being laid carefully on the earth above [59] to increase the area of the primer. A piece of oil cloth — Indian rubber cloth or even tarred paper or oiled paper put over the top of this primer will ensure it going off even in rainy weather.

The primer (**G**) can be made water proof by being dipped in a solution of gutta percha in chloroform repeatedly and a thick coat of tallow put upon its under side renders it effective for a while without other cover from wet.

A mattock, grubbing hoe, or carpenter's adz (**N**), the blade a little straightened, or a spade to dig the holes for the shells are all instruments required in planting them. To transport these subterra shells within the enemy's lines, from a pole supported on the shoulders of two men suspend the shells in a bag — a 10-inch shell is sufficiently heavy (115 lbs loaded) for two. And to carry them in the field where troops are operating they should always be attached to the cavalry and for this purpose two 10-inch shells or others in proportion can readily be carried by a mule one in each end of a bag and it thrown across his back with a pad tied on a pack saddle if he have one.

If the cavalry have artillery — the shells in the caisson will answer all purposes by putting two primers side by side in the eye of the shell, without any plug stopper whatever (see plate 21 [22]).

[**60**] To check the advance of an army along a road, plant these shells in their front, but attach to each sensitive primer a string a foot long tied with 3 loops around the shank just under the head and upon the end of that piece of red flannel about as wide and long as the finger. When the shell is planted this little piece of red flannel is brought to the surface so as to identify the spot where the shell is. Besides this a note is to be taken of the locality of each shell planted as follows "on main road to A.B. in three miles of the city in the middle of the road opposite a large pine tree with two blazes (one black) on left side of the way, are two 10 inch shells." This precaution is necessary in case the shells have to be removed.

Ordinarily the primers are not inserted until necessary on proximity of the enemy and they should never be left unguarded by a sentinel on horseback, until the enemy appears. This is a general rule. As few men as possible should be employed in burying these shells — an officer, a sergeant, and two men are sufficient for this purpose — to lead the mules or drive the wagon (with shells) or caisson & bury them — the officer, sergeant and one man to be well mounted on horseback also the 3rd man if not driving. The sergeant should carry in a haversack — a screwdriver to extract the screws from the holes in the [**61**] fuse plugs , a box of sensitive primers, a few fuse plugs — a gimlet knife & a small saw to make them in case of need, and some disks of oil cloth 9 inches in diameter to shield set primers from the rain. These 3 men should be of tried courage and have a cavalry guard at all time when putting down shells. In planting the shells the man should kneel on the right knee to operate as before stated.

These shells without any connection explode others in their neighborhood, primed with the sensitive primer. How far their influence extends is not yet determined, but supposed to be about as many feet distant as the number of inches in the diameter of the shells.

A shell exploded in a box tears that to pieces, throwing outwardly with great violence the sides in fragments, and this force can be made to communicate, and fire another by means of a wire attached to the farther side, the other end thereof entering the box with the second shell & being fastened to a loose board to pull against the sensitive primer therein.

This 2nd shell can be made to fire a 3rd and so on to any number. The shells are turned in their boxes so as to bring the sensitive primers inclined downwards 90° or more with a perpendicular line, connecting from the top. These boxes of shells & wires buried in the road or route [62] of an army will be marched over by the troops until the whole be upon them, and then the last one is reached, that may be set off by being trodden upon or man observing the movements at a safe distance by means of a magneto electric battery covered wire, & a magneto electric sensitive primer (see plate 11, fig. 1). Thus an army may be destroyed in a moment of time at the will of one man.

To insure secrecy in this matter — a piece of road in the route of the hostile force must be selected and a guard set at both ends thereof, if possible out of view of the operatives planting the shells and after these are all ready the guard must only be withdrawn when the enemy are appearing.

In planting torpedoes in water or land a necessity for concealment has been thoroughly provided. Those intended for water service should be out of sight, or boxed up and labeled another article to deceive. For transport of subterra shells no such precaution is necessary. In fact our object is not to

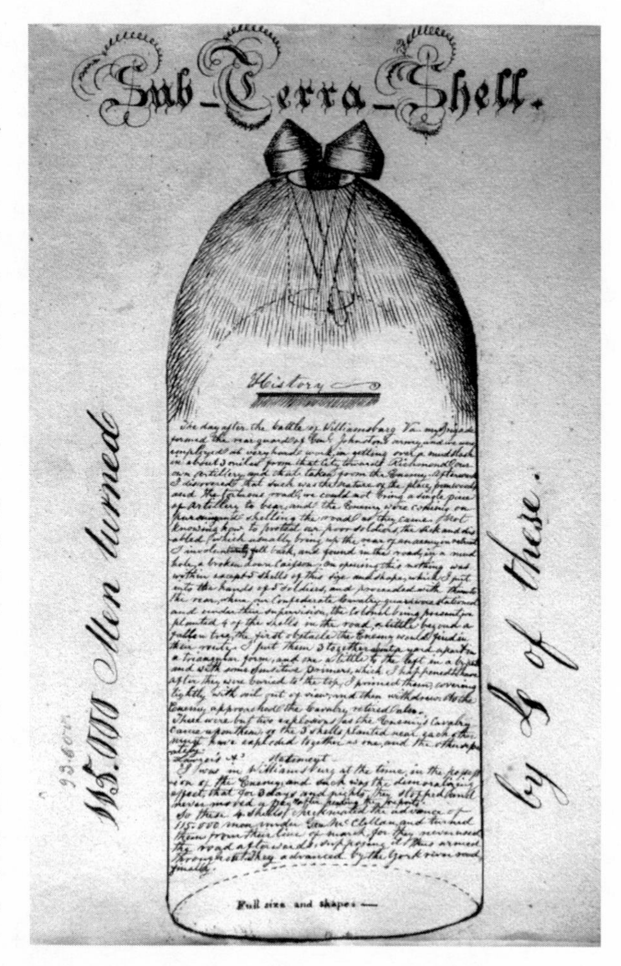

Plate 22 text
[along the margins] 115,000 men turned by 4 of these [above 115,000 in pencil is 93,000]
History

The day after the battle of Williamsburg, Va., my brigade formed the rear guard of Genl. Johnston's army, and we were employed at very hard work, in getting over a mud slosh in about 3 miles from that city toward Richmond our own artillery, and that taken from the enemy. Afterward I discovered that such was the nature of the place, from woods and the tortuous road, we could not bring a single piece of artillery to bear, and the enemy were coming on pursuing and shelling the road as they came. Not knowing how to protect our good soldiers, the sick and disabled, which usually bring up the rear of an army in retreat, I involuntarily fell back and found in the road, in a mud hole a broken down caisson. On opening this, nothing was within except 5 shells of this size and shape, which I put into the hands of 5 soldiers, and proceeded with them to the rear, where our Confederate cavalry guard were stationed and under their supervision, the colonel being present we planted 4 of the shells in the road a little beyond a fallen tree, the first obstacle the enemy would find in their route. I put them 3 together about a yard apart in a triangular form, and one a little to the left in a basket and with some sensitive primers, which I happened to have, after they were buried to the tops, I primed them, covering lightly with soil out of view, and then withdrew. As the enemy approached the cavalry retired also.

There were two explosions as the enemy's cavalry came upon them, so the 3 shells planted near each other must have exploded together as one, and the other separately.

Lawyer's A's statement—"I was in Williamsburg at the time in the possession of the enemy, and such was the demoralizing effect, that for 3 days and nights they stopped and never moved a peg after hearing the reports." So these 4 shells checkmated the advance of 115,000 men under Gen. McClellan and turned them from their line of march, for they never used the road afterward, supposing it thus armed though they advanced by the York River road finally.

kill but to demoralize which a knowledge of this intended use of them is sure to do after the first exhibition of their effects. Subterra shells prepared with the sensitive fuse plug copper disks, as that in plate 5, fig. 2, can be used to protect a landing even on a sea beat shore — or to obstruct a ford over a stream and in fact everywhere the ordinary [63] subterra shell can be used.

To prevent military movements on a railroad plant a shell between the crossties out of sight, and on top of the earth over the shell, put a piece of wood, a stick, chip, stone or anything of the kind, so that as the steam car passes the inner flange of the wheel will come upon it & press it down upon the sensitive primer and explode it. This is a work of but a few moments, & may be done on a road in rear of the enemy by a raid. Such use of these shells, however, is to be made with caution, for fear of destruction of women, children, & non-combatants & should never be done except on roads used for military purposes.

To hunt up and remove subterra shells can be prevented easily & in a way to insure death to those so employed which will be developed hereafter.

For the surprise of a camp by night, ordinarily the public highways must be used,— and the most efficient picket guard is a sentinel and subterra shell or shells planted therein. On the approach of the enemy, the sentinel is to make his escape as secretly as possible using the shell to give the alarm, and to a certain extent demoralize. This is much better than the chance of losing a piece of artillery, which otherwise must be employed for a signal, together with the gunners [illegible] [64] and probably a company of infantry stationed there for its protection.

The fact of one man and a shell on each approaches from the direction of an enemy securing a camp from night surprise, is sufficient of itself alone without other advantages, to establish this arm of service as an essential adjunct to an army in the field.

Troops may retreat, and the bravest quail before a mighty host, but these shells never fail, a force which yields to no circumstances — they are ever ready to resist, & however great the number of the enemy they are sure to find the carnage greater, and as an ever watchful sentinel they are certain to give the signal of alarm at the proper time in thunder tones too loud to be mistaken.

In attempts to approach a camp by an enemy at night, no surer guard can be formed than one, two, or three of these subterra shells buried in the roads of approach with a sentinel in watchful distance. When the foe comes upon them, in their loud report no mistake can be made — the garrison or camp is at once aroused while the enemy becomes demoralized from their mishap.

[65]

Dart Grenades

Plate 13 [14]

They are made as indicated in the drawing of a cast iron shell (A) having a hollow (C) which is to be filled with fine gunpowder (1¾ ounces).

Within the orifice at (I) a tin tube (G) soldered is fixed with lead so that the sensitive primer (F) with its shank can be readily passed in and held in position. Consequently the tube (G) should be slightly conical at the end (G).

For further security the under surface of the primer (F) has a thick coat of tallow, so that when forced in by the fingers grasping that by its cylindrical surface, it effectually closes that opening from wet.

Another opening in the shell at (**H**) is closed by the end of the shaft (**B**) being fitted in tight after having glue just around it.

Diagonally through the opposite edge of the shaft (**B**) the vanes (**B'**) are passed and glued, and then are afterwards greased. Simultaneous with their insertion the strap (**D**) is also inserted after the end has been dipped in a solution of glue and then the string (**E**) is wrapped around that and the shaft.

[66] After the invention of the hand grenade (to which this is an improvement) soldiers were selected to use them of fine form and athletic, and these were called grenadiers. The weapon ceased to be used except in special cases, probably from the danger of using them with fire to set them off, yet the cognomen of this class of persons still remains as applied to the leading company of a battalion or regiment—where the men, until the Florida War, were selected for their height & fine appearance.

The efficiency of this weapon on shipboard none can doubt who have witnessed their effects, for they would clear an adversary's decks in a few moments if thrown upon it.

They burst on striking upon the sensitive primer, when thrown by the hand, into from 21 to 23 pieces, and as each piece flies off with killing force, they become a formidable weapon when hurled among a thick body of troops.

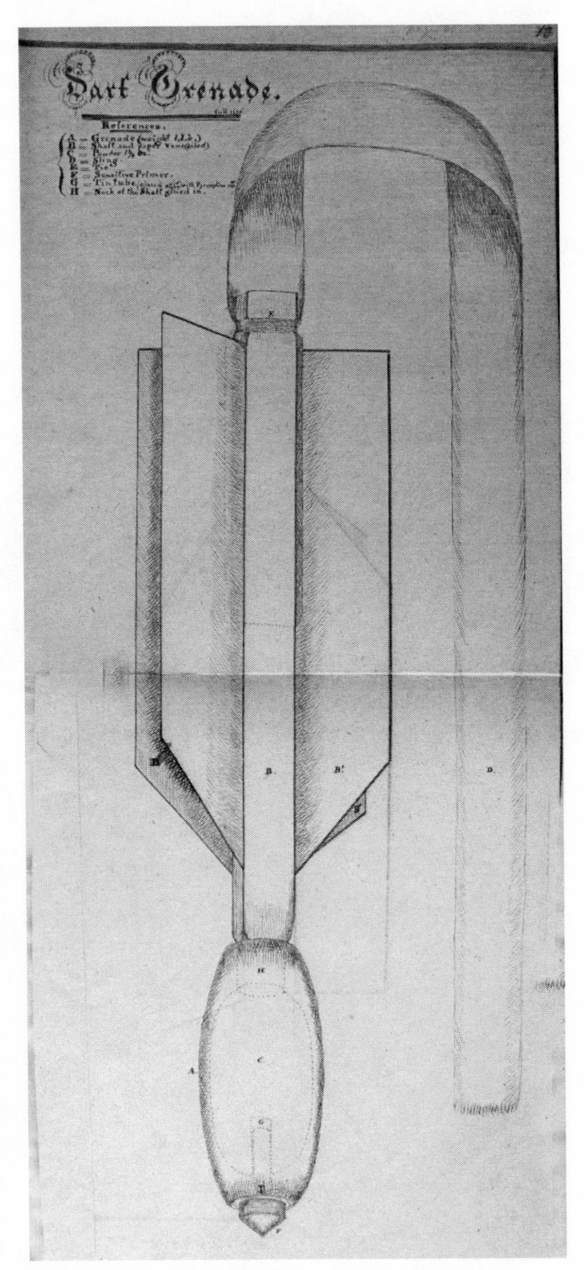

Plate 14: References: ***A***—*grenade (weight 1 lb.)* ***B***—*shaft and paper vane (oiled)* ***C***—*powder 1¾ oz.* ***D***—*sling* ***E***—*tie* ***F***—*sensitive primer* ***G***—*tin tube* ***H***—*neck of the shaft glued in*

A company of soldiers armed with dart grenades would certainly be far more effective than armed with any other weapon, and the facility of using them comparatively free from danger from fire, renders it probable that they will thus be called into requisition again for service in the field and the companies of grenadiers again revised.

[67] To defend forts—earthworks or reveted works there is nothing equal to the dart grenade at close quarters. The strap can be cut off and the weapon thrown by the hand grasping the iron part when necessary & this can be done with more accuracy by troops accustomed to their use, though of course the distance is much shortened thereby.

These dart grenades have been taken into the field in a belt (plate 14 [15]) which gives to a man wearing one properly filled 14 fires, and as each one bursts into 20 or more pieces — one man becomes equal (except for simultaneous discharge) to 280 muskets. The men thus armed have usually accompanied cavalry, and by breaking off the tail part of each grenade, they can be made into very effective subterra shells with the sensitive primers in the tin box and for this purpose by planting 3 or 4 together. The men on duty planting subterra shells are armed with a belt of these dart grenades, each having a due supply of sensitive primers in his tin box, slipped on to the belt.

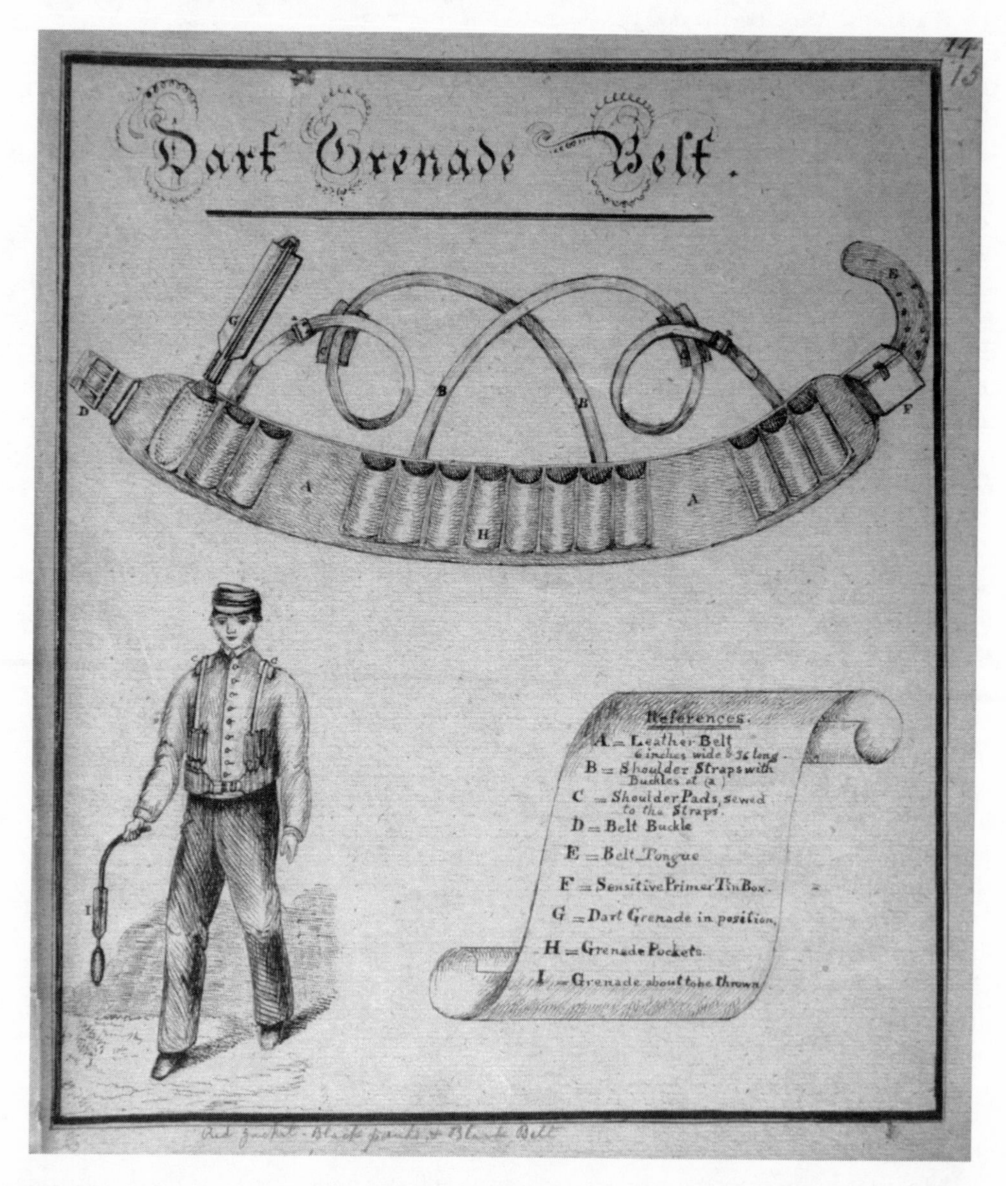

Plate 15: References: ***A****— leather belt 6 inches wide & 36 long* ***B****— shoulder straps with buckles at* ***(a) C****— shoulder pads, sewed to the straps* ***D****— belt buckle* ***E****— belt tongue* ***F****— sensitive primer tin box* ***G****— dart grenade in position* ***H****— grenade pockets* ***I****— grenade about to be thrown*

[in pencil below frame] red jacket black pants & black belt

With these and a few 12 pdr. shells, a small body of cavalry are fully equal to a much superior force as we have found in the field.

To carry properly into effect subterra shells, and for using these dart grenades, men must **[68]** be assigned to such duties exclusively. (The last Confederate Congress became convinced of this and passed an act establishing & organizing a proper [Torpedo] Bureau and Corps,—with 6 millions of dollars appropriation.)

Cavalry ordinarily must confine themselves to roads in moving through a country, and to destroy them with subterra shells—an officer in charge of a party of men should be sent to plant from the center of the road a bed of these shells, then leaving a sentinel there mounted with his horses' head turned from the shells. In juxtaposition have another bed of shells planted, and a similar sentinel also mounted but all of these last from the center of the road to the opposite side—so that cavalry can pass around the horses' heads, making a figure like the letter "S" around the two sentinels & can act as a decoy.

These are to retire when pursued back around each sentinel, who as the squadron passes fall in with their rear & thus leave the road & the two beds of shells to destroy the attacking enemy which the velocity of their horses will ensure. The sensitive primers are left for use in boxes packed with cotton, and this rule is not to be departed from, those in the tin box on the belt are not excepted.

To destroy single individuals is not the object of this invention; where human life is concerned we can **[69]** not be too careful—we should only look to make an impression upon the masses, and thereby render war too horrible to be followed. And if we attain our point by demoralization, it is better then death. There may be some exceptions to this rule, as the death of a victorious leader might paralyze an enemy in the field—or the magazine of a fort hastily evacuated might be blown up by shell & primer put behind the door which in opening strikes the latter, and a railroad car be demolished to prevent their [*sic*] use for warlike purposes; however, ordinarily, the use of these shells for such purposes have been refused—lest thereby noncombatants, as women & children, be killed. To render abortive efforts to find & remove subterra shells put in the route of an army, where there is a probability they will be hunted for, the shell must be placed some foot or two below the surface of the earth over the destructive primer, a stick to be placed upright thereon, the end scooped out to go over & the earth fitted in, so that the feet of those hunting for these shall press upon the top of the same concealed just below the surface. This plan will destroy those out on this hazardous undertaking & thereby prevent it. Without betraying this to the enemy, he may be told that subterra shells are buried this way & where they are, yet he will be frustrated **[70]** in removing them, as they are too deeply buried to be found. This plan must be studiously concealed, that it reach not the ears of either friend or foe, & should be known only to the engineer officers superintending the location of the shells. In fact if these inventions become universally known they must stop all wars, and as far as known, they render superior to all other powers that one, which possesses these.

To Stop Raids

Let 3 men, on swift horses, ride after & round them, and plant small subterra shells in their route—or for some few miles from a city threatened, on the public roads leading thereto, an officer, a sergeant, a corporal & one private are to be stationed, with a few subterra shells already planted, but not primed until assuranced that the enemy are coming that way. Then they can prime the shells, & retire leaving one to guard, to keep friends off, until the actual appearance of the foe, when the guard can also leave on horseback.

All approaches should be thus guarded & a few lessons given will undoubtedly prevent raiders by demoralizing the heads of the leading columns. It has been found that troops will no more pass over roads thus viewed than they will over other land supposed to be mined. They forsake the road not knowing but that there may be many places thus guarded in it.

[71] *A river open to friends, but closed to enemies*

In plate 15 [16] fig. 1

We have the plan of a river protected by torpedoes from ironclad rams and other war vessels of the enemy.

At (C) is one or more submarine mortar batteries made after the plan exhibited in

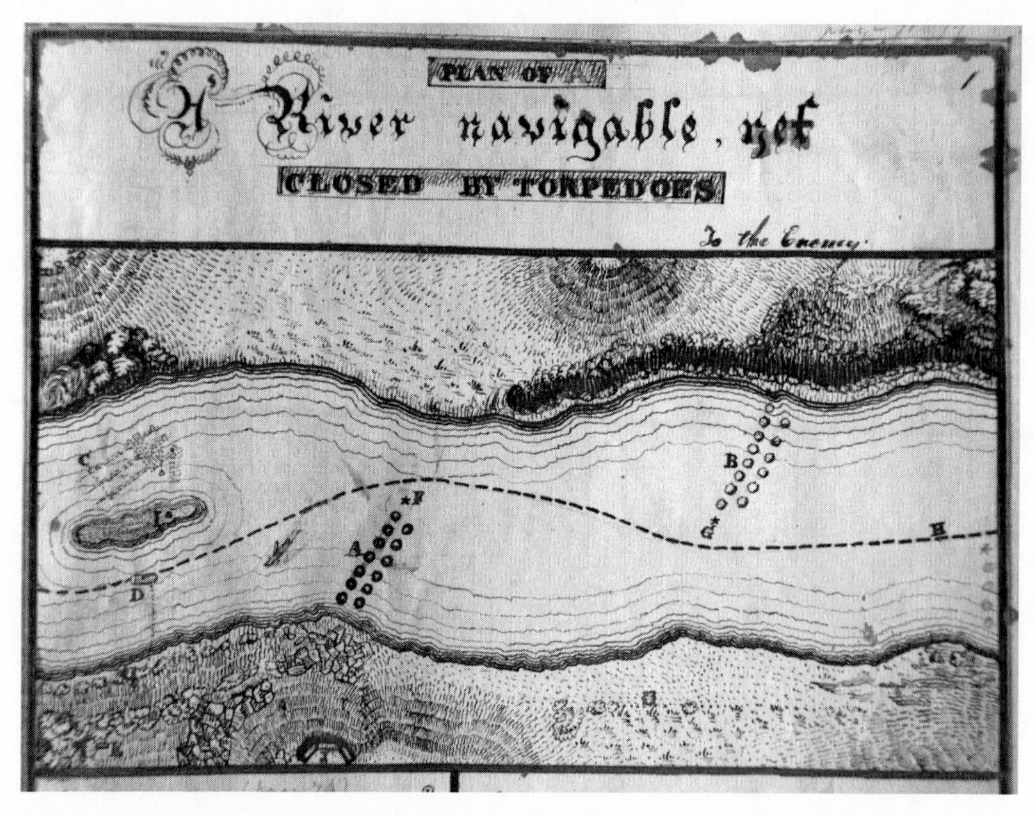

Plate 16, detail of figure 1: References: ***A****—13 torpedoes planted from the center of the channel towards the right bank of the river* ***B****—14 torpedoes planted from the center of the channel towards the left bank of the river* ***C****—S[ubmarine] mortar battery or batteries closing the channel that side of the island* ***D****— magneto-electric torpedo with a charge of gun powder (in pounds) proportioned to the square of the depth of the water in feet* ***E****— magneto-electric battery and the insulated wire leading thereto* ***G & F****—channel stakes around which vessels pass safely, which to be removed on approach of the enemy* ***H****—path of safety—this might be set between 2 stakes of buoys if necessary with torpedoes on each side* ***I****—indicating objects of position of the enemy's vessel by camera or from the point (****E****) when over the mine or torpedo* ***J****—covering battery*

plate 3, fig. 1—which after many months submerged have been found thoroughly efficacious to destroy ironclads or vessels passing over them.

The channel between the island and mainland is effectually blocked up as represented, and any roadstead can be thus protected, provided it be in range of fire from a battery as at (**J**). On the south side of the island a sunk electric battery as in (figure 1 plate 11) to be exploded by magnetic electricity is located at (**D**) with the electromagnetic machine concealed at (**E**), or out of range of fire from the water.

By this the channel can be left open for friends or blockade runners, and closed perfectly to enemies at all times—if they know of it being there, information usually being given by deserters.

Channel stakes or buoys at (**F**) and (**G**) plainly indicate how vessels are to steer in passing the bed of torpedoes at (**A**) and (**B**) and when these are removed, there is nothing to direct an enemy [72] ascending the stream, except objects on shore which ordinarily make it too hazardous to attempt with a small opening between two channel stakes & torpedoes on both sides of the passage can be safely passed and thus removal ensures destruction to an adventurer, who has the boldness to attempt it.

Where it is not necessary that the channel should be left open, a few torpedoes scattered about at uncertain places will have a wonderful effect in demoralizing an advancing fleet of the enemy. A map of the river or harbor showing the position of each torpedo is necessary for future reference and also for examination into the condition of each after submergence during a long period.

The torpedoes usually planted some two or three feet under the surface of the water only can often be examined when the tide is out and moon in apogee, for they are apt then to appear on the surface at a very low tide. To protect a river effectively will often require some 100 torpedoes of from 40 to 60 lbs. of gunpowder each.

[73]

From [manuscript] page 48

Plate 11½ [12]

An electric torpedo to protect the channel can be made by housing within it a glass or insulating vessel to the top of which is connected the insulated electric wire leading to a magneto electric machine on shore, and from this glass within the torpedo (or in a vessel or box at bottom communicating) connected with a metallic pendulum much like a bell and its clapper—to be operated upon. In fact a bell is better than glass with a hole in its top for the insulated wire to pass through & be connected with its clapper thus insulated and containing a sensitive primer as in plate 11, fig. 2, on the outside of the bell. It is operated on this (plate 11½ [12]).

A passing vessel's bottom strikes the torpedo (**A**) and leans it down, when the clapper (**B**) comes in contact by gravity with the bell fastened with metallic fastenings to the torpedo staff, and as the circuit is now completed a sentinel on shore turning his machine sends a spark of electricity which passing through the primer (**P**) into the clapper forming an electric circuit by contact with the bell explodes it & thereby the torpedo.

As the connection for electricity is not complete unless the torpedo leans, so the torpedo does not explode unless in contact with the vessel's bottom to lean it. This torpedo can be used to keep open a river to friends but closes it to an enemy but in strong currents the

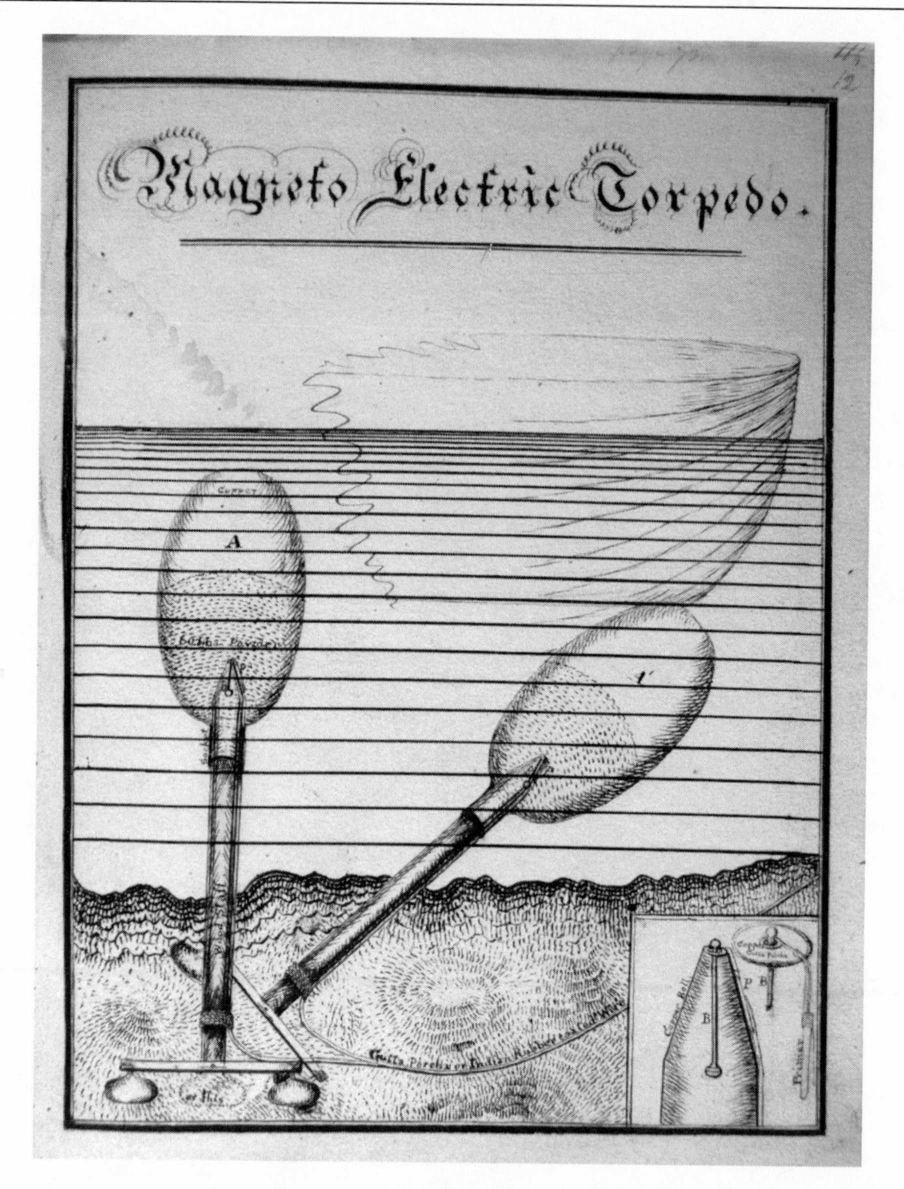

Plate 12

torpedo had better have two anchors, one up stream and the other down stream (see plate 1 where the middle cable (**F**) is necessary) or be stepped as in the drawing but a single mushroom anchor is better.

[74] Incendiary Bullet
Plate 15, fig. 2

This bullet looks like a common rifle slug with its cartridge appended as usual and on a cursory examination it will not appear otherwise. It will fire as well from a rifle and to a casual observer is nothing more than the usual load for that weapon. Yet it is as designated — a most infernal contrivance to do evil, and is a thing not to be known among common people, lest incendiarism fill the land with terror and dread.

The object of these pages is to secure peace by immunity from war, by rendering the latter too horrible to be followed. Devils & damned spirits only could have put it first into the mind of man to take advantage of his fellows by slaughter, and often for such little cause. That cause it as we may, in the last mighty and bloody contest — a vast majority of the troops had not agreed, and actually did not know at last what they were fighting for. These bullets explode on time, so that one dropped into a bombshell & the latter fired into an enemy' camp might lie there for hours or be collected & piled with others when at an unexpected moment it will explode to the destruction and dismay of those around. A man with a number of these bullets, which he could prepare when wanted for use, might go through a city [75] amidst the cotton factories with them in his pocket and imperceived might throw them in places where at midnight they would create a fire.

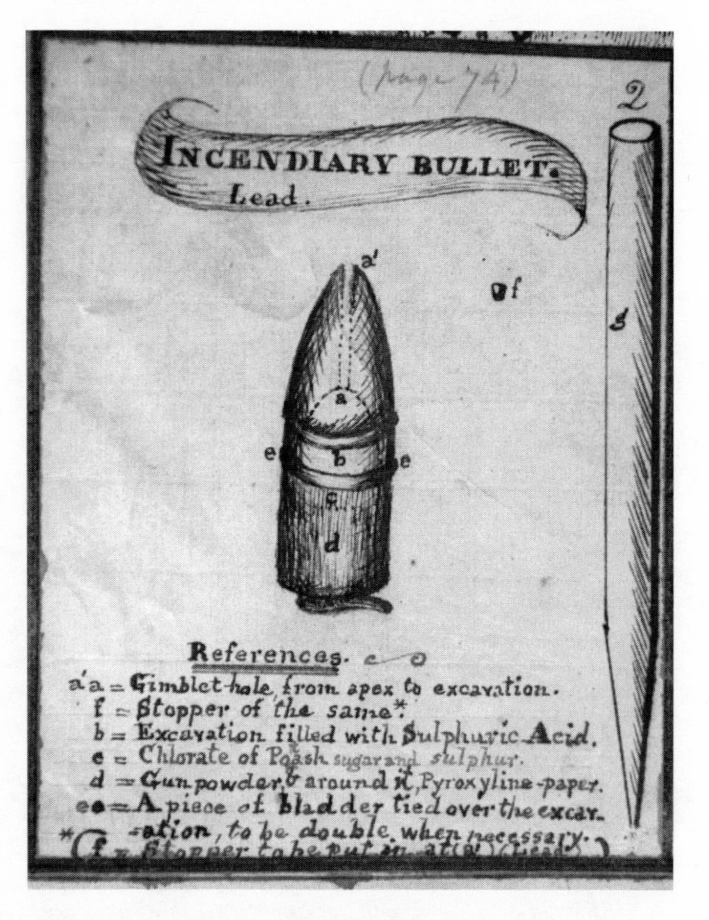

Plate 16, detail of figure 2: Incendiary bullet references: ***a'a****— gimblet hole, from apex to excavation* ***f****— stopper of the same* ***b****— excavation filled with sulphuric acid* ***c****— chlorate of potash, sugar, and sulphur* ***d****— gunpowder & around it pyroxlline paper* ***ee****— a piece of bladder tied over the excavation, to be double when necessary* ***f****— stopper to be put in at (****a'****) (lead)*

To make the bullets, with a small gimlet bore a hole from the apex (**a'**) to (**a**) the excavation, then tie a piece of bladder [**ee**] like a drum head over the excavation say for 5 hours & double for longer times or treble as the case may require (to be determined by experiment). Now form with paper a cartridge to be tied around at (**cc**) with thread and after it has dried pass in a roll of paroxyline paper ½ inch wide and 10 inches long to fit close around the exterior paper shall allowing space within to put in a composition of powdered chlorate of potash & white sugar about in equal parts and then after to be filled up with fine gunpowder & complete as a cartridge. The stopper (**f**) also made of lead to fit in (a') which is to be inserted after sulphuric acid has been introduced and then smoothed down with a knife blade, or something like, to prevent leak & it being observed that there is a hole there.

The sulphuric acid is taken up in a small cylinder (**g**) of glass by dripping the tapered end into a phial of the same and placing the thumb on top of the opposite end, then putting that taper end into the hole of the bullet at the apex & withdrawing the bulb of the thumb when [76] the acid will immediately be discharged into the excavation of the bullet.

When the animal membrane has been perforated or "eat through" by the acid which will occur in a few hours it will fire the chlorate of potash & sugar and that the gunpowder in the cartridge which will burst it & liberate the pyroxyline paper on fire in a blaze.

If a longer time be desire of course a double or treble thickness of the bladder drum head will accomplish it.

A bullet or two thus prepared can be dropped within a shell just before firing & when the shell falls within the lines of the enemy (of course the shell has no fuse but a fictitious one to plug it up) it will explode at the proper time & may set off its fellows if piled up & do no little damage. A few such shells might not only prevent their being picked up but actually might drive an enemy out of his camp, the soldiers not daring to molest them.

*As acid dissolves wire or metal & also mercury &c this [illegible] can be made use of for longer periods

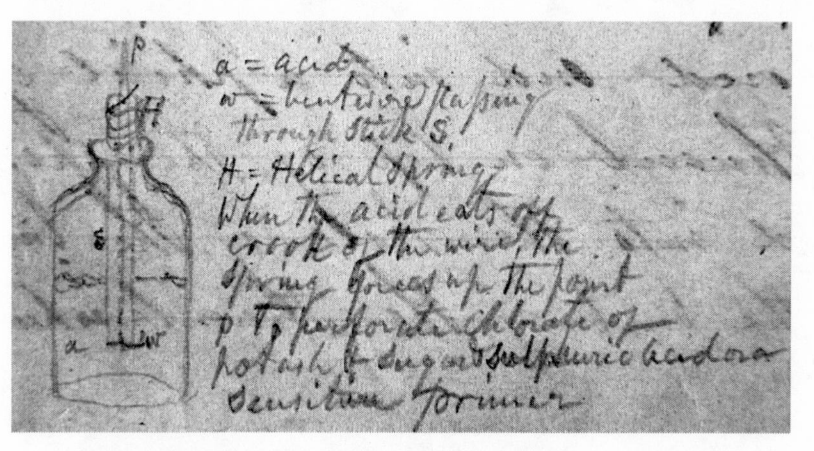

Legend: ***a*** *= acid* ***w*** *= bent wire, passing through stick* ***s.*** ***H*** *= helical spring. When the acid eats off crook of the wire, the spring forces up the point* ***p*** *to perforate chlorate of potash & sugar or sulphuric acid on a sensitive primer.*

[77] The enemy under Genl. [William T.] Sherman checkmated at Jackson, Miss., by subterra shells.

Having been ordered by President Davis to repair to the assistance of Genl. Joe Johnston with my subterra shells, I joined him at Jackson, Miss., and offered my services. The General did not use this arm until on the eve of his evacuation he sent for me and told me that he was going to leave Jackson that night and asked if I could not check the advance of the enemy (under Genl. Sherman) by the shells. I informed him that I could, when he directed me to plant the shells on the city side of Pearl River though I assured him on the coming of the enemy, the women & children would naturally flow that way, it being in rear of the town, & would be destroyed by the shells. I left the general's quarters & sent for Col. [James P.(?)] Parker whom I informed of the general's wishes, but I directed him to have the shells, the largest we could get, planted on the opposite bank of the river to the city, & in the two roads leading therefrom. The colonel obeyed my order, but also those of the Genl. Yet I was happy to find the last the women made known to the enemy & they removed them — not so the others, for their cavalry coming upon them, they opened like a masked battery — threw a horse bodily in the top of a tree, put them to flight, & the infantry hearing the firing & seeing the cavalry retreating, supposed that a masked battery sure enough, so they retreated also. This road to Brandon was not used afterwards & the enemy stopped pursuing Johnston — part of whose army already had retreated 40 miles thus the 2nd time the shells saved our general.

[78]

To Make Sensitive Primers

Fulminate of Silver

Dissolve about 32 grains of ordinary silver, i.e., such as usually is made into spoons or other articles or money, in two ounces of ordinary nitric acid in a retort, by heating it.

Now heat four ounces of alcohol in another retort to the boiling point and away from the flame pour into this the dissolved silver solution cautiously. Nitric oxide will escape abundantly at first but will soon cease, and chemical action in the formation of the fulminate of silver commences. The heat engendered by the chemical action is usually sufficient to continue it until the whole mass appears in crystals. But should it not, cautiously heat it a little again but avoid the puffs as much as possible as they probably are exploding crystals from too much heat or action.

The crystals are of peculiar shape — long and tapered arrow points truncated at the base.

If mixed when moist with fulminating mercury, though made the same way, it is decomposed, or becomes no longer sensitive.

These crystals of cyanide of silver can be put on a paper & that scorched before the fire without exploding, yet the smallest friction with sand [79] will set them off leaving an impression that it is electricity which does it, as it is extremely sensitive to such influence (better not be touched, however wet, with a silver spoon). A ball with some of this salt upon it presented to an electric jar charged, so as to draw a spark, its explosion is so similar that the whole appears as one mighty spark.

Mix this in quantity only for seven primers at a time while still wet with powdered glass or sand (washed in nitric acid and water — then in pure water previously) in rather more than its bulk of the wet sand. **(N)**

Fulminating Mercury

This salt is made in the same way as the last with very little variation and yet they cannot exist together, & it would be an interesting experiment to see how for the mixture or amalgam of the two metals would be affected in its formation.

Mix this in *bulk with gunpowder — mealed in equal parts **(M)***

Required also gunpowder granulated (rifle powder)

Gum Arabic in solution. Bees wax

Gutta percha dissolved in chloroform

Thick writing paper, white tissue & cartridge paper

Paste-board and wooden topped cylinders for primers, turned to about 11/20 of an inch in diameter and 1/4 in length

[80]

Manipulation
Plates 16 [17] and 17 [18]

With the punch cutter (**a**) plate (16) [17] cut out of thick paste board placed upon the end of a short vertical log by means of a hammer (**b**) a number of disks and with the smallest punch cutter (**c**) make a hole in the center of each as in plate (17) [18] fig. (**z**).

With the shears (**d**) & paper pattern from a number of sheets of cartridge paper placed on top of each other cut out the forms (**y**) and roll each round the former (**g**) and with gum Arabic "paste" on the inner surface of the sloping side make a tube (**s**) or rather a number of them to put aside to dry.

With the punch cutter (**e**) cut out disks (**x**), and with the shears cut from the circumference to the center as indicated in the figure, and of these by a lap make cones (**w**) with the gum paste which set aside to dry, these to be of thick white writing paper. Also out of thin silk paper with the punch (**a**) cut out disks (**v**) and with the shears or a punch, disks

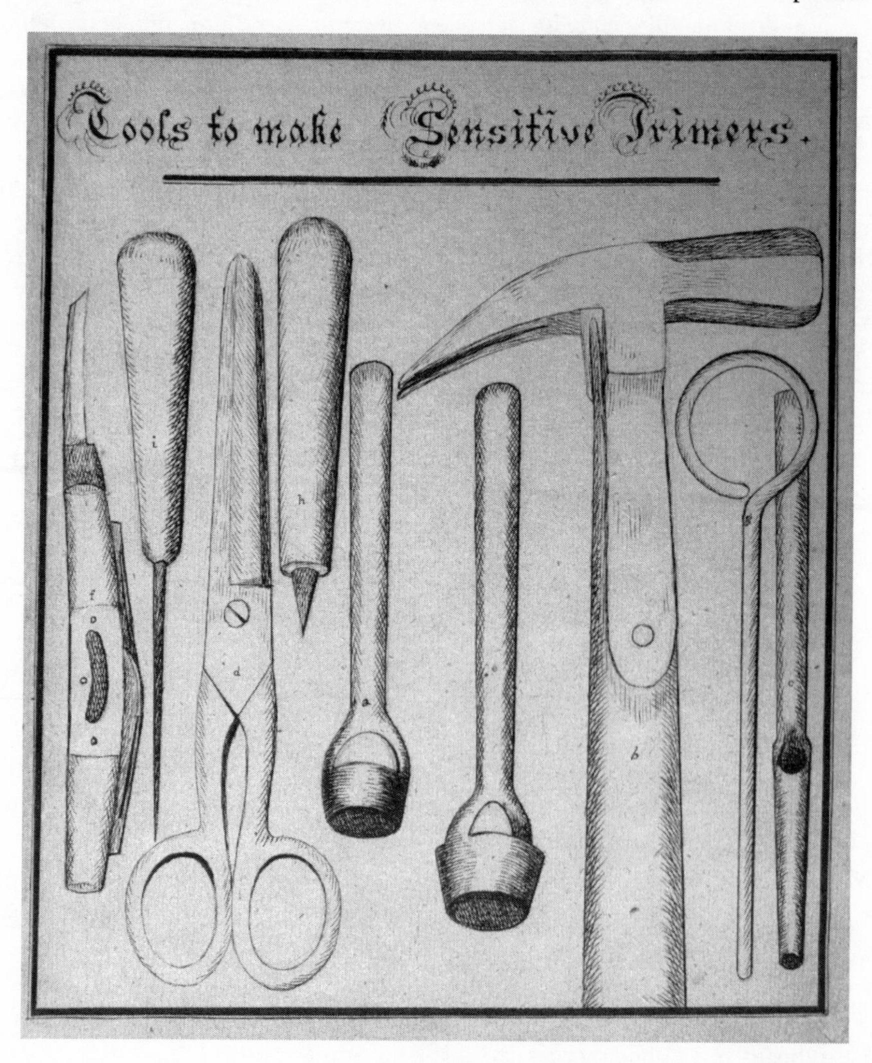

Plate 17

(**u**). After the tubes (**s**) are dry, pass each through the hole in the disk (**z**) as at (**t**) and with the knife (**f**) cut the top down to the disk as indicated — spread out with the finger and with tool (**h**) flatten down to the disk after wetting with gum water & put a disk of paper (**v**) upon it & set aside to dry by sticking each into the hole of a primer board (**r**) which is ¾ inch thick, with paper paste for the lower side to stand the primer upon.

[**81**] After the tube is dry, fill it with rifle powder at the lower end & stop up with bees wax, then invert & with awl (**i**) force a hole through the cap-paper and into the powder two inches and fill up this excavation with fulminating composition (**M**) see [manuscript] page (79) and let it come over the top as (**q**) and (**p**) but so as to leave a slight margin around the disk exposed to be gummed, and tissue disk paper (**v**) pressed down upon it, & the primer set aside to dry in another board (**r**) after which it is ready for the final operation.

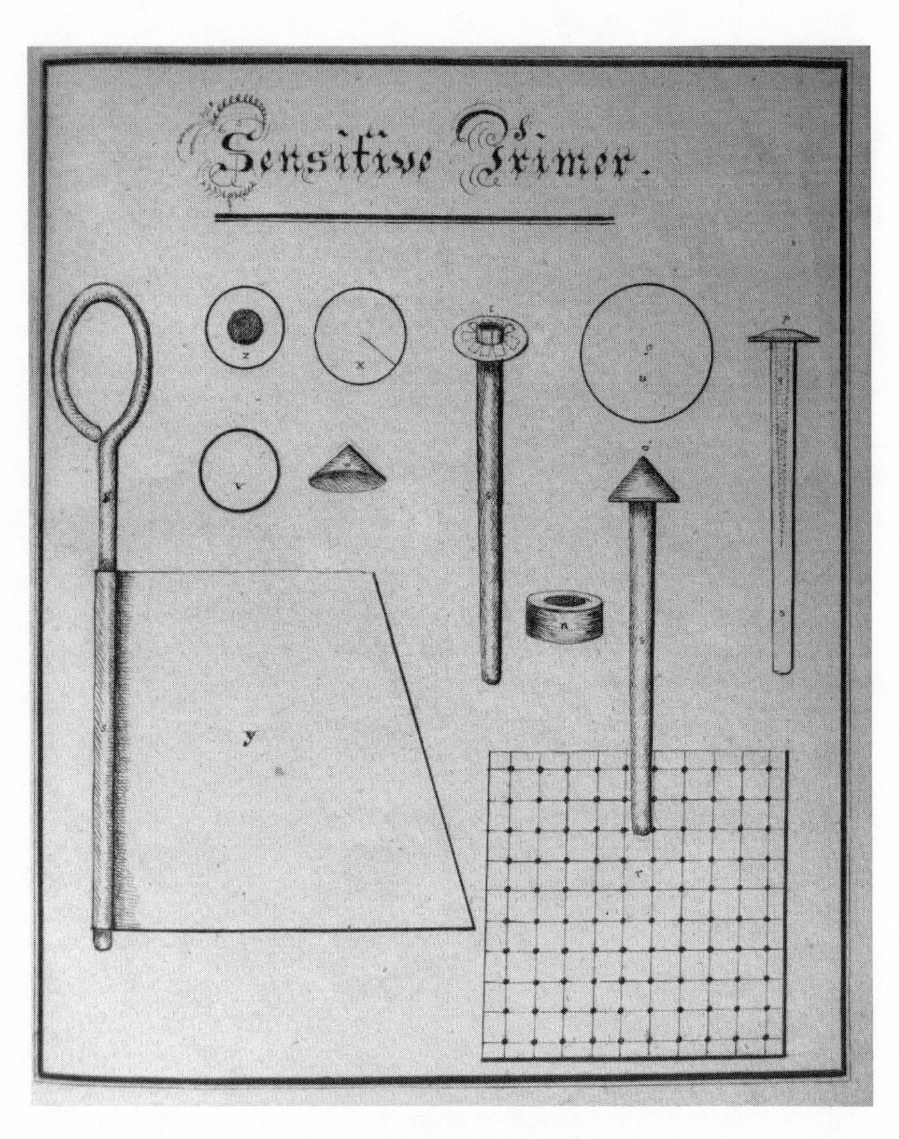

Plate 18

Now holding cone (**w**) inverted by the thumb — the fore & middle finger of the left hand, fill it cautiously with wet composition (**N**) & inverting primer (**q**) place top (**p**) upon the composition in the cone then turning the whole upright & wetting tissue disk (**u**) in gum water place its center (**o**) upon the cone top (**o'**) & bring it down upon it & around the edge under the bottom of the disk where it is to be cemented. (I have usually used a blade of a knife for this purpose.)

After the sensitive primer has again become dry, dip the cone end in ink, then dry then again the second & third time and finally into thick gum water, or paste, to finish it. If intended to be water proof it may be dipped quickly into gutta percha dissolved in chloroform; this may be repeated as often as necessary.

Where a wooden disk (**n**) is used, it is only necessary to dip the paper tube (**s**) in glue before passing it into the disk.

See plate (1) figures (2) and (3).

[**82**] A quill tube (small size) is considered preferable to paper & can be made the same way applying pressure to keep down the nicked edges upon the disk until dry. On plate 12 [13] fig. (**G**) is delineated a guarded or protected primer, by means of a tin cap (**a**) fitting over the top & wooden cylinder where it rests with a tight fit & paper cemented around it, but so as to allow a tread to force the tin cover down upon the primer to explode it.

These tin covers or shields can be made rapidly by means of a pair of scissors to cut the tin, a blowpipe to fuse soft solder laid with a little resin or rosin within the short cylinder & a candle lighted also a file to trim around the base.

To prepare Electric Sensitive Fuse Primers
See [manuscript] page (50)

Prepare a number of chicken or small size turkey quills and cut them of suitable length — plate (11), fig. (2) one of which is shown marked (**E**) in the diagram also two wires as at (**A**) and fasten these together with silk thread by passing it over and between such alternately bring them very close to each other with the ends projecting as indicated.

Now through a hole made in the "pith" of the quill pass these two wires thus united within as from (**H**) to (**a**) and bring the two ends (**d**) and (**e**) round the edge & up over the side where they are held in position by a silk thread passed around.

In the end (**y**) open, put in the following components:

[**83**] 1st equal quantities of fulminating silver & charcoal finely dusted and commingled, in quantity sufficient to reach above the point of the two wires to (**x**) then fulminating mercury and mealed gunpowder also mixed & in equal quantities up to (**z**) then fine gun powder & coarser to the small end (**y**) which is to be plugged up with wax. Finally dip carefully the end (**H**) into a solution of gutta percha in chloroform, just so as to form an imperceptible coating & when this is dry, dip again up to (**b**) & ultimately the whole tube if necessary to render it waterproof.

Retort carbon can be used for the above and this method of primer has been found superior to the plumbaginous mark on cork or paper, being more sensitive to electricity. The very small spark from an ordinary medical magneto electric machine passing through the human body sufficing to explode the primer. (This is also a discovery of my own, but whether it is the same as that of Wheatstone in England or something superior I cannot say as that is unknown, but this tells for itself both for mining purposes and torpedoes.)

It has been found that a tin tube to enclose the primer adds much to its efficiency, in fact it seems that the fulminating silver set on fire by the charcoal acts not by its fire, which is too quick, but by its percussion force, much like the hammer of a fowling piece upon the copper cap on the nipple [**84**] and a metallic tube to confine its gasses plays the part of the copper in the cap.

It is thus made:

Make a tin tube ¼ inch in diameter and 3 inches long (I make them round a former with a vice and hammer, and solder by means of a blowpipe & rosin). Now the quill tube (E) being ready plate 11, fig. 2—one of the wires (**d**) is bent inclined upward and carried around the top of the quill for future use. The other wire (**d'**) is also bent at right angles so as to leave the quill tube free to wrap paper round & round it to make it fit tight when inserted in the tin tube (**t**) which is now to be done (It has been found that when exploded that part of the tin tube opposite the fulminates bursts with a loud noise, so on the outside of this, put another tin tube about 1 inch long & solder, though, this is not essential). Now having covered the end of (**d'**) (½ of it being cut off) with solder (by brightening and working it in melted solder & rosin)—bring this down to the tin tube & solder it on. This can safely be done if held in a cold vice, by a copper solder soldering iron, so called. A copper thimble (z') is next soldered on to make electric communication with the tin tube easy. From (**m**) to (**n**) now wrap paper colored blue or otherwise & from (**q**) to (**n**) with red over the blue. Use thread around the blue paper to make it fit tight until it has dried from the paste & then remove the red paper. To fire the primer case fully pull out the wire (**I**) and connect with the wire of the magnetic electric battery.

[85x] David's [*sic*] or Torpedo-Boat

At Charleston, Mobile, and Richmond, a number of small boats from 25 to 30 feet long made of boiler iron plates with a locomotion engine which operated a propeller at the stern as the motive, were made and called torpedo-boats (fig. **A** plate 19) or 'Davids.'

These boats carried on the end of a spar a torpedo as arranged in a larger scale in (fig. **B**) with points of steel, so as to strike an enemy's vessel below the water-line and stick in the spikes (**c**) when on backing the boat it would be left and explode by electricity and a protected gutta percha covered wire, or a cord pulling down a slab upon a sensitive primer.

It was one of these boats which struck the [*New*] *Ironsides* at Charleston [October 5, 1863] with a torpedo in front, in command of Lt. [William T.] Glassell of gunboat *Charleston* but though it failed to destroy, it rendered her inoperative for a very long period thereafter.[20] (The torpedo was fired by concussion and the reaction so terrible that the wave of water swept back over her & down her smoke pipe or chimney nearly putting out the fires, sweep off the lieutenant who was taken prisoner next morning by the enemy while hanging onto an anchor chain (see page [manuscript] 101).

The engineer of the boat contrived to get up the fire again with another man on board, escaped back to Charleston. These boats can also be effective by carrying torpedoes to be pulled floating against the vessel's bottom by means of rope—as indicated in the diagram. They will not admit a spar if sufficient length in front to be out of danger with a gunpowder torpedo, but with eleven pounds of guncotton equal to 60 lbs [**86**] of powder, compressed to a cubic foot, the spar might be 40 or 50 feet long making it safe from a percussion exploding torpedo, and can be useful or made so in many ways (20 feet of water between

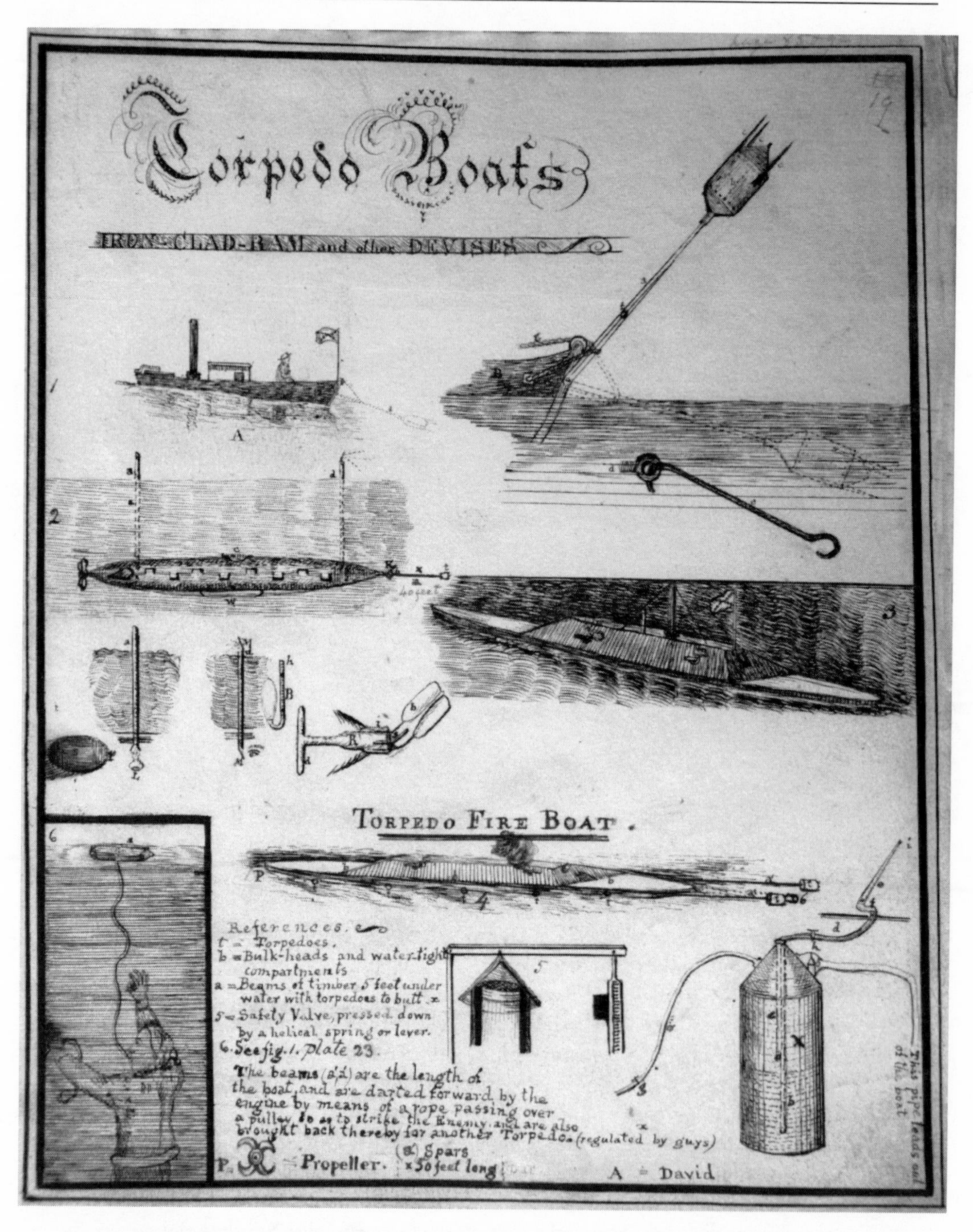

Plate 19: References: ***t****— torpedoes* ***b****— bulk-heads and watertight compartments* ***a****— beams of timber 5 feet under water with torpedoes to butt* ***5****— safety valve, pressed down by a helical spring or lever* ***6****— see fig. 1 plate 23*

The beams ***(a'a')*** *are the length of the boat, and are darted forward by the engine by means of a rope passing over a pulley so as to strike the enemy and are also brought back thereby for another torpedo (regulated by guys)* ***a'*** *spars x50 feet long* this boat & the torpedo is considered safe). The torpedo (**b**) is lowered into the water by the rope (**g**) and can be secured in proper position by means of side ropes or the iron stanchion (**e**) hooking at one end into the bolt (**d**) which screws into the hole near **B** and the other end into iron staples fastened into the spar (**a**) at (**f**).

Electricity to explode the torpedo at the proper time should be engendered by a magneto electric machine as as [*sic*] noted in plate (11) (or by a sensitive primer the explosion could be caused). The crew should consist of but 3 men *viz.*, the captain to steer & the engineer and a ready man.

It is easy to make the small cabin shot proof against small arms so as to approach an enemy's vessel & receive with impunity a sentinel's fire — and in a dark night to act energetically, & then retire, her small size rendering her too small a target for larger guns to hit. 5 of these Davids will conquer any ironclad.

By so arranging the spar and its hinge at the head of the boat, as to allow it being turned over so as to lie lengthwise of the deck, it will admit of greater length,— and the torpedo at the end may be so buoyed up by the "vacancy" & gunpowder within, as to be of little weight upon the bow. So arranged it will be able to butt an adversary with impunity & destroy her thus with her torpedo provided the spar be so fixed as to be immediately detached afterwards. This boat, nearly submerged, might be made to navigate by steam and electricity without a crew to guide it by wires from shore.

[87] Iron Submerged Propeller or

Torpedo cigar boat (plate 18 [19], fig. 2) an improved plan, one that noted under the head of torpedo boats. A description of this boat is as follows.

Length 25 feet. Color black. Diameter vertical 6½ feet. Diameter horizontal 5 feet. Motive power 8 men working cranks and one to steer — seated within the cranks at right angles alternatively and working a propeller at the end on the outside. The steering apparatus two fins, one on each side, moveable as inclined planes without and by handles within (**x**) or by the rudder as detached enlarged marked (**R**). The rudder at the opposite extremity from the propeller of the boat made in the shape indicated (fig. **R**) in proper position as at the point (**K**) to handle (**d**) working inside the boat turning the cylinder (**R**) working through a stuffing box so as to allow pushing out or pulling in and turning to the right or left.

The blade (**b**) is fastened to a spring (**s**) ordinarily but in the manner indicated but assumes a straight position on being pulled in the surrounding cylinder (**i**). It will be perceived that by pulling in or pushing out and turning the handles (**d**) to the right or left any direction can be given the boat whatever. The boat should be ballasted and have an iron or leaden disk (**w**) of such weight and secured by a staple passing within, that it can be detached at any moment by a blow, and suffered to fall off when the boat will at once rise to the surface, the hatchway (**c**) be uppermost (see page [manuscript] 106).

There shall be a water barometer (**B**) whose end (**h**) **[88]** should so pass through the top of the boat to allow the water to press upon a column of mercury in the tube (**B**) the air in the glass bulb yielding thereto so as to indicate the pressure and consequent depth of water without on a graduated scale. The tube need not extend beyond the upper surface of the boat but should be conspicuous to the steersman at all time.

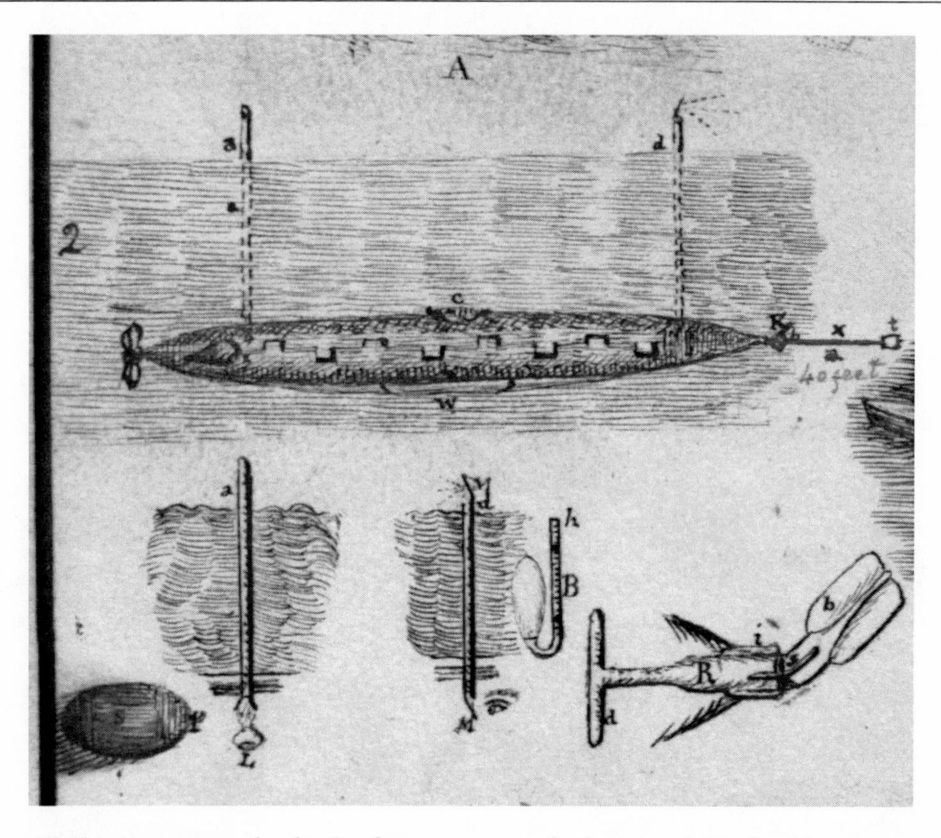

By this instrument the depth of water can ever be known which should never exceed a depth to be ascertained by experiment.

By having a box "centrally" open under the bottom of the boat with a ¾ inch rope to wind on a shaft, the crank handle to be operated inside the boat the weight (**w**) can be used as an anchor & the boat anchored in the position indicated. (**a**) is a tube of India rubber with a spiral steel spring so arranged with cords as to be capable of reduction in length and by it the lamp (**L**) which lights up the boat within keeps up a draught by the heated air and tube (**d**) which communicates with the upper air & has two mirrors (**M**) and (**M'**) inclined each at an angle of 45° so that an eye at (**M'**) can see on the surface — can supply the necessary draught downwards of air. This tube must be made water tight, to slide up & down & turn around so as to see everywhere & be capable of reduction in its length. By means of these two tubes about 1½ inches in diameter each, a circulation of air can be kept up as to enable the crew to remain any length of time below they might wish, lying concealed below trees or bushes growing from the banks of a river.

[**89**] This boat can be made to navigate by steam by causing the fire inside the boiler to be blown by bellows — the exit of the draught to be by a pipe (**a**) — This plan has been found very effectual. For safety the rope attached to weight (**w**) might have a knife to cut it at any moment operated by a wire passed within. To make this machine useful, torpedoes with coils of rope, attached to the outside might be so fixed as to be liberated from the inside to be drawn under, or against a hostile vessel.

The figure marked (**S**) is a soda fount-like vessel, two or three or more which with compressed air might be carried within for safety. The boat should be so arranged that by means of a force pump water could be let in below the floor & forced out when necessary controlling the specific gravity thereof. This space below the floor with the two ends might be made to communicate and a force pump operated within to inject water so as to compress the air within about ten atmospheres affording both a supply of vital air & means to sink or rise. A cock communicating with the cabin to obstruct & let in the air, and one to let out the water below.

It is believed that light might be obtained enough by a magneto electric machine for this torpedo boat with ½ dozen Bull's eyes without a candle or lamp, which should not be used to vitiate the air by burning up the oxygen, if it can be prevented.

Air can be compressed to ⅛ of its bulk (120 lb. to square inch) by a hydraulic force pump pumping in water in cell **H** the force pump to be operated by the [illegible] until the vessel sinks to the required depth, then by opening a cock in the bottom the air will again expand and the vessel will then rise as the water is driven out by the air. By reversing the position of the cells, a diving bell can be used on this plan.

[90] Fire Boat or Incendiary Rams

(In plate 18 [19], fig. 3) will be found a likeness of the Confederate ironclad rams and this of the *Tennessee* in Mobile harbor. The guns so arranged on their traverse circles as to fire either out of the side ports or ends. The iron shield is 5 & 6 inches thick. All such boats should take in water before going into action so as to sink as low as possible—having a force pump so arranged as to discharge it also when required.

As the caliber of guns increase, so must the oblquity of the shield as testudo becomes greater and their height above water less, with increased thickness of shield. In (fig. 4) is a plan for an ironclad-ram & fire boat where the guns fire from ports close to the roof the two guns by means of traverse circles as above firing on each side & in front or rear.

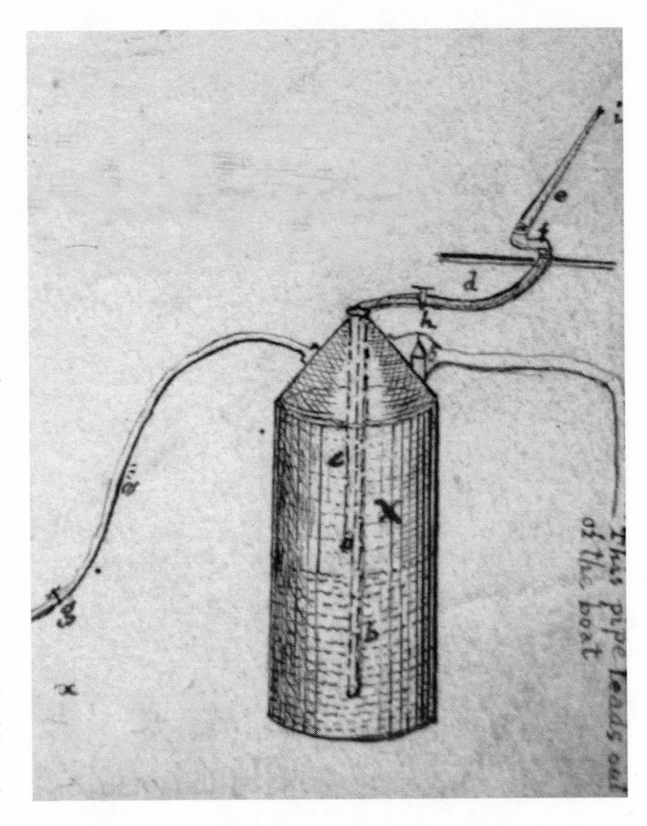

The **X** is a vessel containing sprits of turpentine and alcohol or naphtha with an arrangement of pipes within as indicated, *viz.*,—one (**b**) terminating in the spirits and the other (**c**) terminating in the alcohol but both opening into tube (**d**) and communicating with the throwpipe (**e**). This throwpipe manipulated from the upper deck is to have universal motion by means of a gooseneck (**f**) like that of a common fire engine and the tube (**d**) passing through the deck as indicated.

This vessel (**X**) is to be just out

of the way as in some nook or corner but with the pipe or tube (**a'''**) communicating with the steam chamber of the boiler of the boat so that by the stopcock (**g**) the pressure of steam can be let in [**91**] upon the inflammable fluids mentioned or benzene or kerosene, naphthalene &c. Now on turning stopcock (**h**) and applying a flame to the orifice (**i**) a stream of liquid fire can be thrown upon an enemy & into their ports covering with a wave of flame. This boat is to have no chimney but to keep up her fire in the boiler furnace by bellows, a plan found to be very effectual. Her waste steam & smoke is escaping from an orifice in the upper deck as exhibited in the diagram.

This boat is also made to sink down preparatory to action by taking in water & to rise again by forcing it out, which can be done by steam from her boiler.

To prevent being run over she is to have submarine mortar shells (plate 3, fig. M) set around her below the water line on booms. All ironclads to be effective should have a cylindrical iron tube projecting from bow or forecastle, like a steam boiler without ends, say at an angle of 45°, to slide down a torpedo fitted to another tube (telescopic sliding) arranged so as to approach and strike an adversary below her water line to blow her or any number up, see plate 22 [herein labeled 24]. (The torpedo is represented of the largest size)

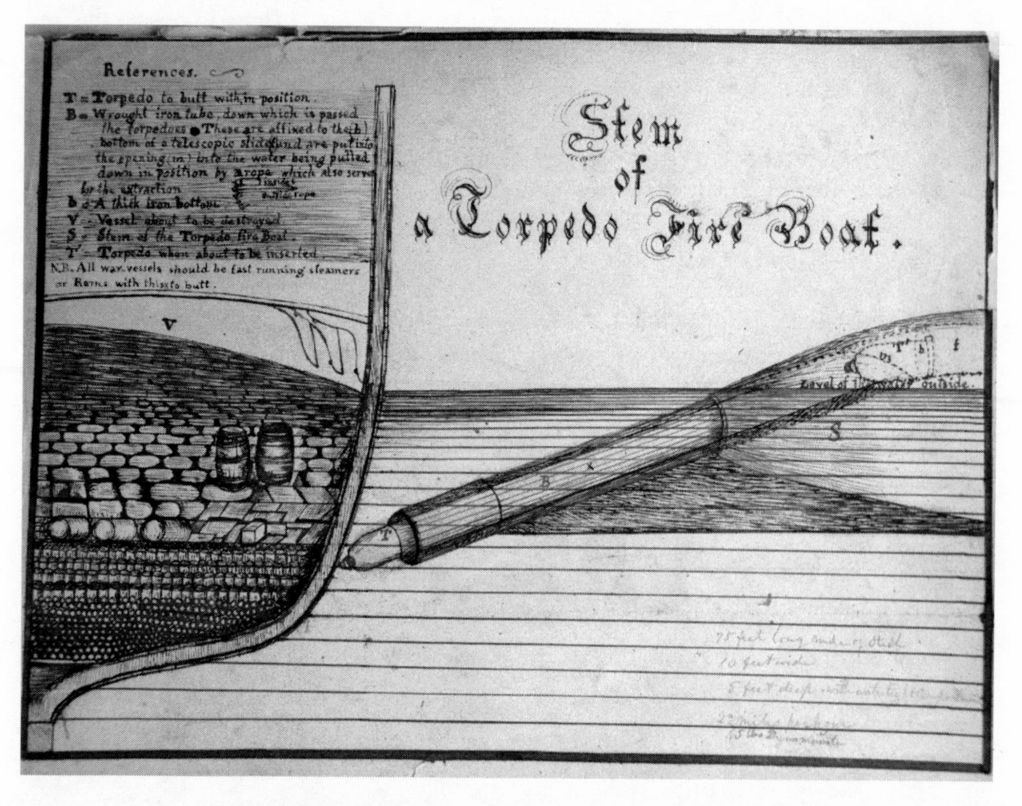

*[Plate 24] See Plate 18 [19] References: **T**— torpedo to butt with in position **B**— wrought iron tube down which is passed the torpedo. These are affixed to the (**b**) bottom of the telescope slide (**f**) and are put into the opening (**m**) into the water being pulled down in position by a rope which also serves for the extraction. **B**— a thick iron bottom **V**— vessel about to be destroyed **S**— stem of the torpedo fire boat **T'**— torpedo when about to be inserted N.B. All war vessels should be fast running steamers or rams with this x to butt*

[in pencil right lower corner] 75 feet long made of steel 10 feet wide 5 feet deep— with water tight compartment 22 miles per hour 45 pounds dynamite

Safety Valve

After various experiments, it is found that the only true safety valve in the world is of the form shown in the diagram,—*viz.*— that of the Chinese cap, fitting over the top of the escape pipe. The elements of this cone must be at a right angle or less—all other valves though called safety valves should be called unsafety valves (see fig. 5 [Plate 19]).
[92] Fluids at rest have a great tendency to join fluids in motion so that the pressure of the atmosphere 15 lbs. to the square inch is exerted more or less to keep down all such valves except this one—and the greater the pressure of steam upon the valve to open it, the more it is resisted. If the angle made by the elements of the cone of this approved valve make an obtuse or one greater then a right angle with each other this will also be pressed down, steam escaping.

A valve with a flat surface will only permit a small portion of the steam to escape; in fact the escaping steam from between the disks will actually force the disk against the top, against a small force exerted to keep it affixed. In another form, the normal one *viz.*, a cup dropped as a conical frustrum inverted, into its seat the power is still greater to hold in its seat the valve, and so much so that it is not strange that boilers burst, but strange why it does not occur more often and this would inevitably occur were it not for the fact that the engineer, uneasy & surprised that the steam does not escape more freely, aids the steam by lifting up the valve from time to time—thereby allowing it to escape. Who has not observed this fact on our western boats? So this as in the diagram is the only real safety valve or one worthy of the name.[21]

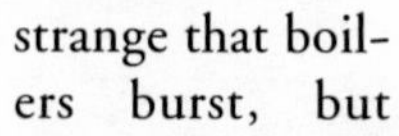

[93] To walk under water

In (figure 6 plate [19]) is represented a man walking at the bottom of a river out of sight except for a floating log (**a**). It has been found that with an Indian rubber or gutta percha tube ½ inch in diameter & 8 or 10 feet long one end in the mouth and the other end reaching above the surface of the water, a man can (if properly weighted down) exist & progress and be made instrumental in attaching rope or device & pass unnoticed a sentinel up on the bank. The log through which his breathing tube passes of course must & be of such a nature as to appear a floating stump or something like so as not to attract attention. In certain cases a knowledge of this fact might be useful [;] he might remain under water some hours.

Deception is the art of war. A fictitious or stuffed straw man held up by a stick and a man below the rest of the parapet at Battery Wagner, Charleston Harbor, would have discouraged the sharp shooters of the assailants by receiving all their shot[s] & made thereby to believe their guns, or aims, defective, which would make them careless in their fire.

[94]

Flight of Projectiles
Plate 19 [20], fig. 1

It is a very mistaken notion which many military men have that a gun fired inclined or even vertically is strained more thereby. On the contrary the stress is less than in any other position, as the inertia is the same at the moment of impact—when the bursting power is greatest, whatever may be the position; and when not increased by friction the shot or shell must leave the gun more freely as the angle of elevation or depression becomes greater.

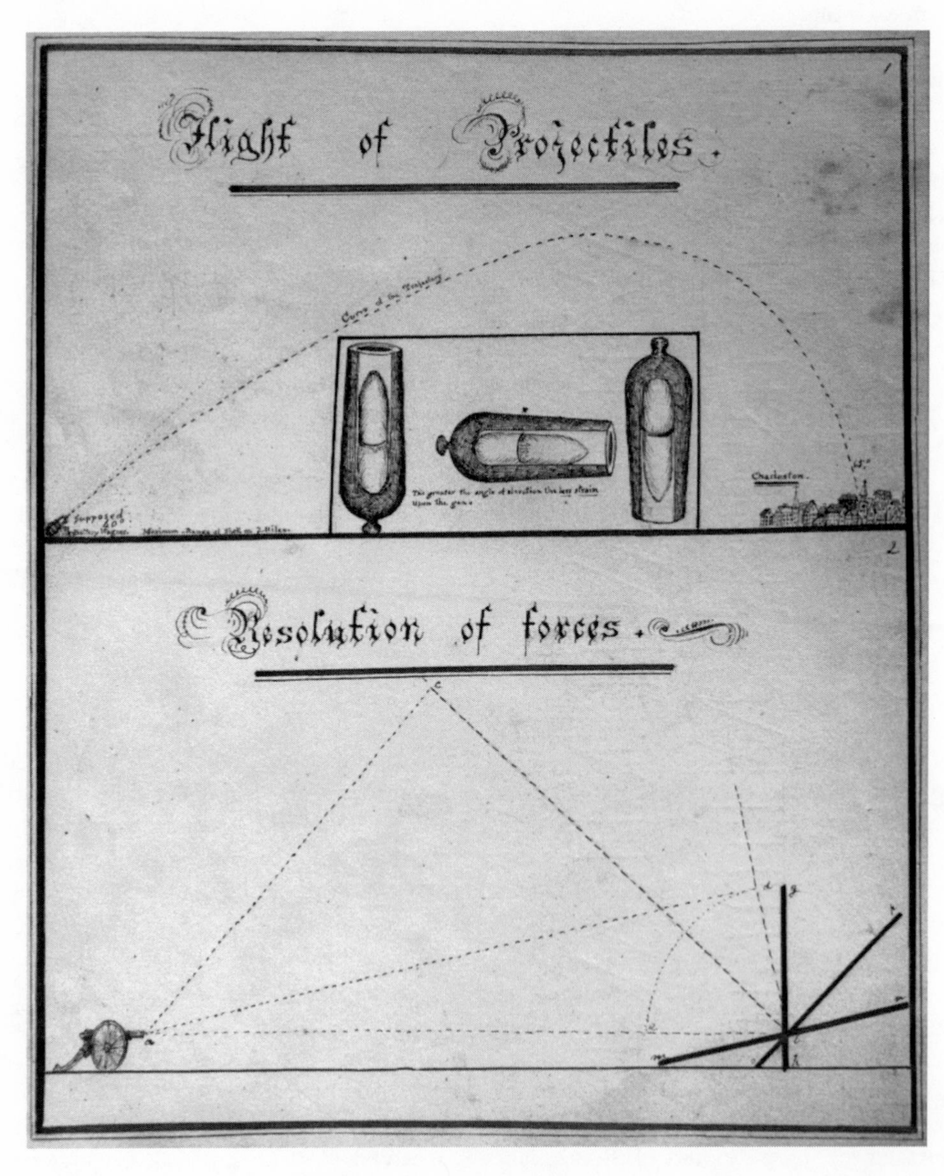

Plate 20

The inertia is as the weight always, but when there is added to that weight the friction of the shell moving along the bore of the piece, increased no little by the developed gasses acting as must needs be the case upon its upper side as in fig. (**x**) plate (y [21]) pressing the shell down against the lower side of the chamber, when the gun ranges horizontally, the stress must be proportionally greater, and the liability to burst increased. Experiment at Augusta Arsenal, Ga., has proved the correctness of this assumption.

Persons have been led into the above error thinking by the weight of the shot or shell, without taking into consideration the very sudden development & vast power of the gasses liberated from fired gunpowder behind it in comparison with which the weight of the shot, aside from its inertia, counts almost nothing, say shot 120 lbs. to more than 33,000 lbs. to square inch. Rifling a gun increases its range greatly and the Whitworth gun with not rifled, but hexagonal bore **[95]** like the cell in the honeycomb of the bee, twisted twice and half around itself, seems to afford the greatest range. This is a breech loading piece. The utmost range attained in the late war was at Charleston, S.C., from Battery Wagner firing into that city — total distance of seven miles. The shell thus sent fell into the city at an angle of 65°, those on the percussion principle did but little damage considering the numbers fired, many of them not bursting, but those with time fuses burst above the city and the pieces of shell passed all over it.

Many of them fired elevated as supposed 40° bursted, attributed to the strain on them by the elevation, whereas the greater the angle of elevation or depression, the less strain as stated. When a gun is horizontal then the cartridge & shot lies on the lower side of the chamber, leaving as necessarily must be a small vacancy or windage between the cartridge, shot, or shell and the upper side. The gasses developed when fired must at once pass into this opening or vacancy and press with immense power downward like a wedge upon the shot forcing it against the lower side and thereby bursting the gun. If the pieces have been loaded some days & moisture in any manner has rusted the shell or projectile on its lower side in position, the danger becomes greater of a burst when fired and the gun is broken as it were over its own shot like over a fulcrum. It is an entire mistake among military men that the greater the angle of elevation or depression the more strain there is upon the gun.

[96] Resolution of Forces

Showing the necessity of inclining the iron shield or testudo the figure in plate [20] fig. 2 is given.

Let the base (**ab**) represent the full force of the piece when discharged against the plane (**hg**) placed vertically.

Incline the plane to the position (**op**) and from the point of impact (see fig. 2) erect a perpendicular (**bc**) to that plane and from the point (**a**) draw the line (**ac**) perpendicular to (**bc**). The force (**ab**) is resolved into the two forces (**ac**) and (**bc**) and (**ac**) represents the effective force upon that plane.

Again incline the plane to the position (**mn**) and erect there from at the point (**b**) the perpendicular (**hd**) and from (**a**) draw the line (**ad**) also perpendicular to the line (**hd**)— the distance or part (**db**) will represent the force of the concussion or impact upon that plane, and (**cb**) its equal when measured off on (**ab**) will indicate what proportion of the original force becomes effective [and] what is lost.

Now the force of impact is equal to the quantity of matter multiplied into the velocity, and even in the extreme case of a obliquity (**mn**) where but about 8/37 of the breaking forces effective, 15 inch shot must destroy ironclads

[97]

Observations

A spar some 100 feet long or more with a torpedo on the end, and in its rear a short board (**b**) or inclined plane to direct it advancing affixed at an angle of 45° or so by turning the spar it will go to the right or left as might be worked from this board though lying in the water to strike an enemy's hull with the torpedo end depressed, suspended by the float, so as to hit below the water line (or see plate 18 [19]).

A long floating beam with a torpedo on the end of the same divided, however, & connected with a universal joint will answer better.

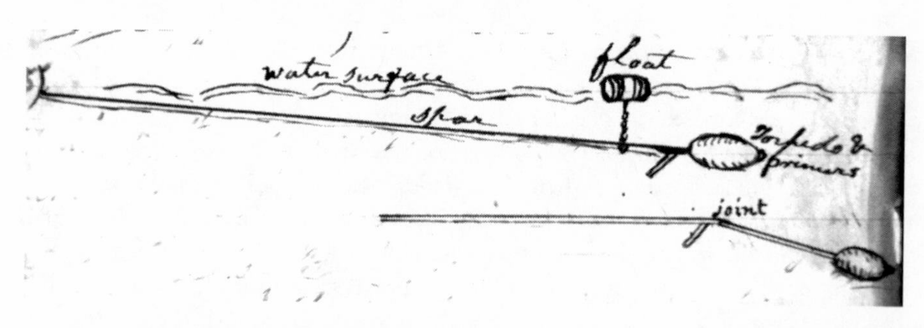

Compressed carbonic acid might be made use of as a power to navigate the air or guide a torpedo boat through the water, though steam is probably better as on [manuscript] page 16.

And the "Fireless Locomotive" *Scientific American* page 118, 18 of August 24th 1872[22]

A bird evidently flies by using his wings, tail, and body at an inclined plane, his flapping being but to overcome the slight resistance of the air or atmosphere after he attains the necessary initial velocity. A buzzard uses such little exertion to sail & fly that man can undoubtedly do it with the proper knowledge & machine.

Nitroglycerine is made by adding successive small quantities of glycerin to a mixture of one part nitric to two of sulphuric acid keeping it cool, & when dissolved poured into water — the amber colored fluid is the article to be exploded by a blow which when mixed with clean sand, looking like coarse brown sugar, forms dynamite which can only be exploded by a percussion fuse. These articles might be made useful for torpedoes.

Gun cotton, though twice as strong as gunpowder, takes up more then twice the space.

[98]

Yorktown [in pencil]
Rains' System of Electric Telegraph — plate 22

Which can be learned by any one in five minutes. For a commander to telegraph his orders it is useful, and in fact whole communities in civil life can profit by it and it thus increases the power of the post office immeasurably to process it.

By it the letter copied itself line by line 160 in a minute though at 1,000 miles distant, and the office requires but a single man, whose duties shall be to read only the direction & person addressed, & to pass it to the manipulator, who then puts it into his machine (**M**) to copy itself at the distant office, 800 syllables or words in a minute. This system requires a key (**K**) on which the letters of the alphabet are marked by corresponding indentions (**I**) like the teeth of a small saw. It is made of brass or other metal and is used to mark the letters with a punch. The first indention begins the letter which occurs most often *viz.* (e). The second (d or t), the third (o), the fourth (i or y), the fifth (s), the sixth (n), the seventh (r), the eight (a), and so on for (l) (c or k), (h), (p or b), (g or j), (u or w), (f or v), (m), (qu), (x or z). It will at once been seen that the intervals between holes are the letters of the alphabet, a mark or two holes consecutively noting the termination of a word by a dash at the distant office. The holes are made by small hand punch (**P**) each sentence beginning with two holes and also every line.

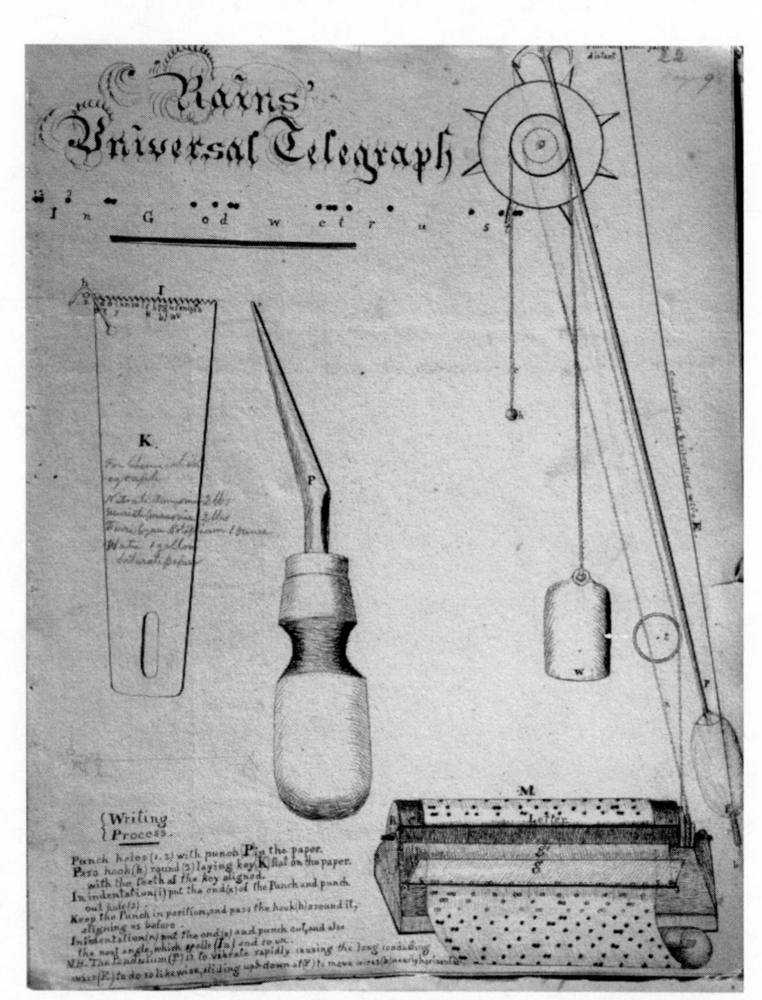

*Plate 22: Writing Process: Punch holes (**1, 2**) with punch (**P**) in the paper. Pass hook (**h**) round (**2**) laying key (**K**) flat on the paper with the teeth of the key aligned. In indention (i) put the end (**x**) of the punch and punch out hole (**s**) Keep the punch in position, and pass the hook (**h**) around it; aligning as before. In indention (**n**) put the end (**x**) and punch out, and also the next angle, which spells (**In**) and so on. NB. The pendulum (**P**) is to vibrate rapidly causing the long conducting wire (**E**) to do so likewise, sliding up & down at (**F**) to move wires (**b**) nearly horizontal*

*[In Pencil on "**K**"] For Chemical Telegraph Nitrate Ammonia 2 lbs Muriate Ammonia 2 lbs Ferricyan Potassium 1 ounce Water 1 gallon Saturate paper*

Process

(see plate 22) Punch two holes at the beginning of every line. The key is laid upon the page. The teeth on a line with one. The ruled copy or letter page and in the last one of the two consecutive holes from punched out to begin a sentence the hook of the key is passed [99] around the punch held in near an upright position and at the required interval the first letter of the sentence, another hole is punched out and so on till the word is spelt ending the last letter with two holes. As the punch is applied to the paper, a slight turn in position is given to cut out the hole. A similar key applied to the received message reveals the letters, which after a little practice can be read without the key easily. By always beginning with the vowels a, e, i, o, u—a transmitted sentence can be easily read key or no key.

A magneto electric machine is to operate this telegraph and the received letters are marked on paper soaked in yellow prussiate of potash, nitrate of soda, and starch. Wm. Bain uses prussiate of potash, nitric acid, and ammonia.

Remarks

Though the (d) might be mistaken for (t) yet on pronouncing, the word is readily recognized, as also (i for y) (c for k) (p for b) (g for j) (n for w) (f for v) & (x for z). The holes are punched out of the sheet of paper (not a strip) and that sheet which has recorded letters or communications is passed under the inclined glasses (**g**) its edge passing into the cut (**C**) so that as the roller (**R**) is turned the sheet of paper is pulled under the glass over a sliver strip (**s**) where the exposed holes are swept by the platina brush (**b**) as the pendulum (**P**) vibrates. Along the pendulum is a wire communicating with the platina brush (**b**) and an insulated fulcrum (**F**) continued in connection to the distant office, the silver slip (**s**) having metallic communication with the damp earth. The weight (**W**) keeps the pendulum in motion & is wound up by pulling on (**h**) the cord (**c**) passing in the cut of the ratchet wheel easily [illegible] by a clock maker. The sheet (**t**) is to keep the cord (**n**) sufficiently tight so as to turn the roller (**R**) & is on the opposite side of the pallate wheel.

[100] Other devices can be used to transmit the words but this, with this machine, is as follows—as the platina brush passes over the holes in the paper it meets the silver strip and the electric circuit is complete when the electricity passing from the magneto electric machine up to the metallic but insulated fulcrum at (**E'**) not shown in the drawing for obvious reasons, but is the upper end of wire (**E**) & down by the wire (**E**) through the brush & silver strip records the mark at the distant station by means of a similar machine, the pendulums each being let loose at the same moment from a fixed magnet by the electric current transmitted and each vibrating in the same time.

The style of this last must be an iron spring point & not a platina brush, however, and it marks upon the prepared receiving paper the holes made in the transmitting letter[s] to be read with a key when received. A little girl 14 years old learned the way thus to communicate thoughts in 5 minutes or less time. The instruction is so easy that one accustomed to make the letters, can write nearly as fast as with a pen. To correct mistakes, punch new holes and fill up those punched out with the small disks & gum arabic. Letters are conventional marks as signs of our ideas. Our ancestors wrote in Old English or German text & it often required many minutes to write (& ornament) the first letter made & was improved by the running hand & Roman letters, so it matters not what the sign is, if it convey our thoughts, & this alphabet is learned easier than any other, even by those who cannot write.

The vibration of the pendulum being in a course back and forth the brush of platinum wires must be so made as to slip in and out of the tube (**r**) so as to sweep the silver lines (**s'**, **s**) horizontally or in a straight line. A second **s'** pendulum would transmit 18,000 syllables in one hour or more words or syllables—48,000 however is computed.

[101] The Valve of a Bunch of Lightwood

During the blockade of Charleston, S.C., a Lieut. [William T.] Glassell of the gunboat *Charleston*, proposed to take charge of a little David, a name we have given to a little propeller steamer, in this case cigar shaped, calculated to carry but four men, and a torpedo at the bow for the purpose of attacking the [*New*] *Ironsides*, a vessel whose daily fire into the city of Charleston gave more annoyance than all the rest of the blockading fleet.

[James H.] Tomb was Engineer of this craft. Walker Cannon pilot, and [James] Sullivan, fireman.

The little boat, which could make some 6 or 8 knots per hour, was about to leave the wharf on which lay some lightwood and a gentlemen present remarked to the engineer Tomb, "You had better take it." "Well, pitch it in," said he, which was done accordingly, and the David departed. She showed but little out of the water except her steering wheel and smoke stack and was of that class which the enemy declared more to be dreaded than a seventy-four gun ship.

The night was dark and the rolling billows not too rough and we may well imagine the feelings of the adventurers as they sped on their course boldly for the mouth of the harbor and the formidable war vessel in front. **[102]** As soon as they approached the *Ironsides* they were hailed "Who comes there?" A shot from a double barrel gun was the answer to the question which left no doubt of the character of the Goliath's assailant, and it was immediately followed by the roll of drums and call to arms in a universal commotion on deck.

Onward sped the Lilliputian steamer, and reached with its torpedo, triple spiked from its bow, driving them in just below the waterline, a little abaft midships & with such impetuosity, as to break off the torpedo staff, as that exploded when its sensitive primers were pressed to the vessel's side. A streak forked like lightening fire with a noise of thunder followed, shaking the mighty fabric to its center, whose total destruction was only prevented by her thick timbered side at the spot where she was struck. A huge wave rolled back upon the little David, and poured down her chimney upon the fire sweeping off the commander Lt. Glassell, who swam to the anchor chain, and firemen Sullivan, who made for the buoy,

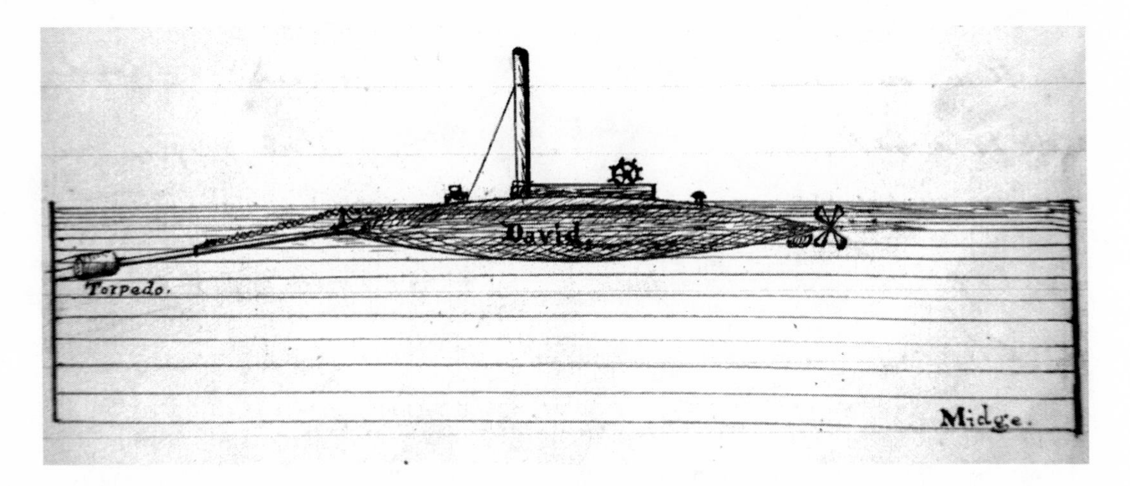

Cigar Boat Midge

from which they were ultimately taken by the enemy. Cannon the pilot & engineer Tomb stuck to the boat & we may well imagine their joy when on turning over rocky ballast from off the drenched fire, some live coals were found, which served with the lightwood immediately to rekindle the fire [**103**] into a blaze, which raised the steam in the boiler still hot, and they were enabled to back out, amidst numberless rifle shots which riddled their smoke stack as they retired from the ship.[23] There was a hail of bullets, but all above their heads, so without being harmed, they returned to port & their wharf in Charleston. Their safety was owing to that bunch of lightwood. Had the David struck the *Ironsides* amid ship she would have gone down, but it was fortunate perhaps for her crew that she exploded her torpedo against a part strongly build, or they would probably have gone down with the ship. The torpedo should have struck & then exploded by lanyard as they were retiring.

[104] Observations

The smoke stack in front should have been formed of two sheet plates meeting at an angle of 60° to ricochet the shot and also a similar shield should have protected the steersman as the rest of the crew might have kept below deck, and he could have piloted by a replicating mirror. This angle by trial I have found will ricochet off shot with boards & and single advancing line of infantry might be thus shielded.

I have found on trial that there is no necessity for a funnel flue at all, as a common blacksmith bellows carried by the machinery is amply sufficient for the draught, without smoke stack or chimney. The quantity of powder in the torpedo was 45 lbs & should have been 60 and it was so near the surface of the water as to act as a petard, it should have been deeper to act as a mine. As 11 lbs. gun cotton compressed in a cubic foot of space is equal to more than 50 or 60 lbs of gunpowder its lightness often recommends its use instead for torpedoes.

[105] To close the Mississippi to an enemy

This was the hardest problem Genl. [Edmund P.]Gaines Comd. Offr at New Orleans [in War of 1812] considered demonstrable and he designed two immense rafts anchored on each shore to meet in the middle of the river at an angle projecting up stream but the great force brought to bear by the current would have caused this to fail & the rafts would have gone to pieces crushed against one another, but a system of torpedoes as represented by the diagram will effect it & yet allow friendly vessels to pass. These torpedoes should explode by contact[;] in the path of friendly vessels the torpedo should be set off by electricity from shore (see plate 24 [23]) and should be so located as to break the interval on each line or wire rope. There should be also two 15 inch guns to command the passing over the electric torpedo.

Plate 23

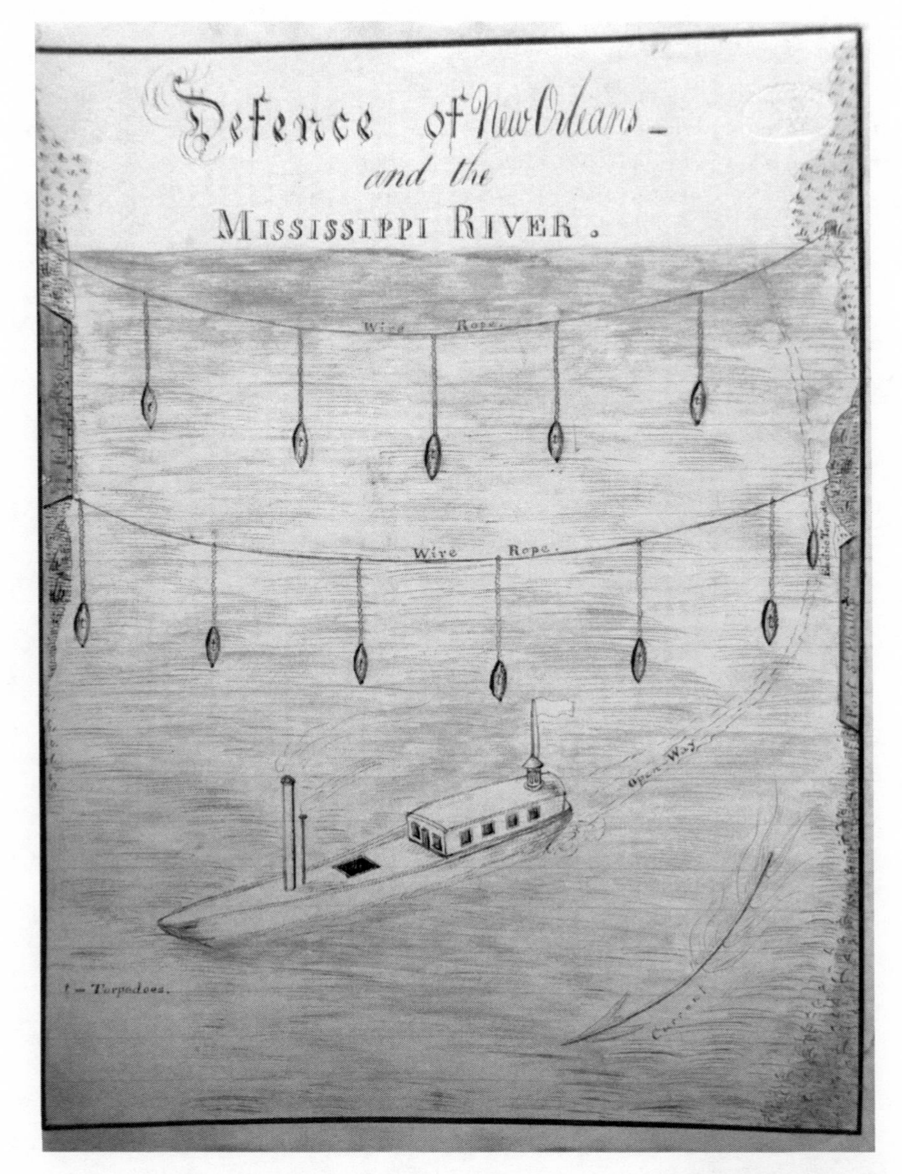

To fish for Torpedoes

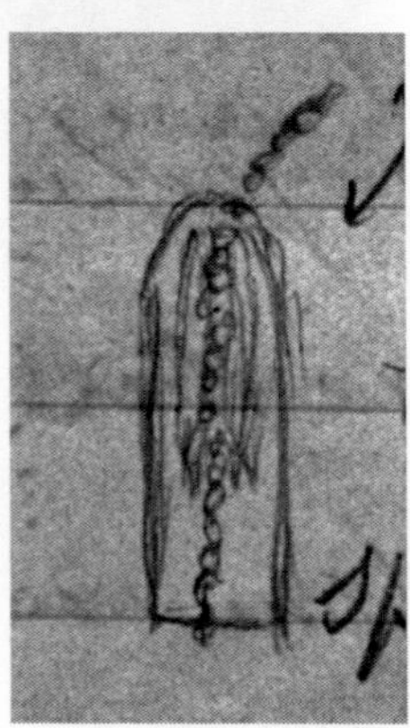

Have a projectile with a flange hinged at top to fit in groove with springs to open them retain in place by a hook which is to be secured[?] and when the projectile is inserted and fastened to the bottom a chain or wire rope to pass up in a groove also & out of the cannon. When fired the flukes wide open & take hold of torpedo wire are being [illegible] & so break it — this machine may be used to anchor ashore in case of need.

[106] A Fire, Super & Submarine Torpedo Boat — plate 23½

This boat is intended to be navigated with or without a crew & with a crew to go under water she should have some soda founts filled with compressed air from 3 to 6 to open with a stop-cock so as to admit fresh air when necessary — with a water barometer, see opposite [manuscript] page **b** or [manuscript] page 18 [88] to indicate the depth of water — a weight fitted on the bottom of the boat symmetrically to be detached from within

Plate 23½

at any moment & to be used as an anchor {to retain at a certain depth the boat see [manuscript] page 84 [87]} and a compartment to be filled & exhausted of water at any time so as to rise or sink the boat. She must have also a steam pipe connected with the boiler of the man of war (**PP**) or if not a steamer, to a separate small steam boiler to heat per caustic soda water so as to get up steam as in a magazine. She must have a spar (**s**) 30 feet long—say 3 joints so as to fold over & be out of the way ordinarily with a torpedo at the end of it, & have 7 or more Bull's eyes (**b'**) & valved to afford light below—or a magneto electric lamp to give light both day and night. From the stern of the boat a projecting cylinder (**w**) have passed through it the three wires to steer the boat to the right or left & make come up or go down—(one wire) and a time machine on board properly fixed would suffice for electrical action—the rudder (**r**) & wings (**q**) being moved thereby. The hatch (**h**) is to be open with a hinge (**?**). She must also have an observation tube as on plate 18 [19] (**mm**) and the liquid fire arrangement as (**X**) and when lying below the surface at anchor, the tube (**a**) with lamp (**L**) might come into play (see [manuscript] page 88). **[107]** A thin & elastic blade is the best propeller for a torpedo boat.

For Ships of War to Meet Ironclads

Each ship must have hung from davits two submarine boats prepared to navigate under the water thus prepared to navigate by steam as the David on [manuscript] page 103 only having neither fire nor smoke stack but a magazine of steam as a boiler full of tubes within a boiler, whose outer shell [is] well protected by felt from losing heat, is to contain caustic soda or other ingredient in excess in water so as to be capable of receiving heat 375° or more, which can be imparted by admitting steam only at 212° from the ship's boiler if she be a steamer, to condense therein. This heated water is to raise steam in the tubular boiler within for the navigation, and the boat is to be guided by a cable containing 3 insulated wires—*viz.*—one to turn to the right or left see [manuscript] page (**p**) or it may be done by a compass needle turning by positive or negative electricity. Another wire is to regulate an inclined rudder to make her ascend and the third wire to stop or go ahead. A fireless locomotive was tried at Patterson, N.J., with wheels 36 inches [in] diameter 7 × 10 inch cylinders. Boiler 37 inches in diameter 9 feet 6 inches long weighing 6 tons with water heated to 150 pounds per square inch of stream which ran 7 miles with a loaded horse cart on a track laid on uncommon road & the remaining steam pressure was 40 pounds.

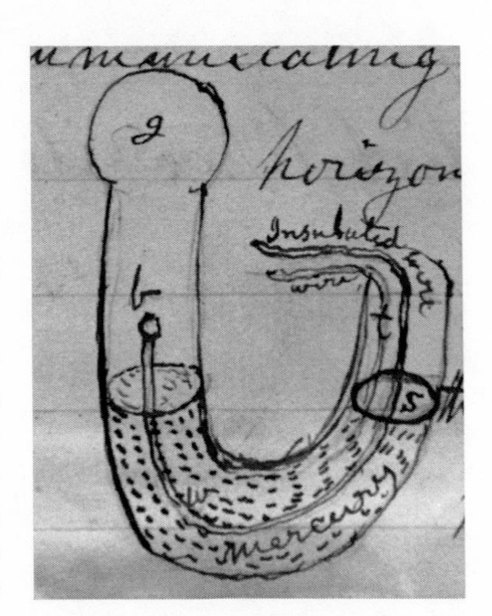

The boat may be made to run habitually at a certain depth under water by having a bent glass tube containing mercury, its open end communicating with the water outside. The inclined rudder being made horizontal by electricity when the mercury reaches the metallic bulb (**b**) by pressure of water on the surface outside the wire insulate the conducting electricity. There is another wire (**t**) communicating with the mercury making <u>that</u> part of the circuit **[108]** as the boat sinks in water the pressure of the water at surface of the mercury (**s**) will be greater and the air in the glass bulb (**g**) will be compressed accordingly until the metallic bulb (**b**) is reached where an electric communication will be made with a battery through its insulated wire and the steering board will be raised & the boat raised. A glass tube thus bent with mercury in it to tell the depth of water should always be used in all boats navigating below the surface of the water, a want of this probably was the cause of the submarine boat on page (**b**) sticking her head in the mud under the school ship in Charleston harbor as she was found.

Observations

A torpedo launch with dimensions — length 1. Beam 1/7. Draft 1/27 Propeller velocity 22 miles per hour is superior for war (see page 343 *Scientific American* of May 30th, 1874).[24]

Torpedoes to be effective must explode by pressure & not by percussion and must be in contact with the vessel's bottom when exploded. A fact which is the secret of my success. Nations only know the effect of torpedoes through me & this principle seems to be neglected generally.

When a torpedo explodes under vessel's bottom the water from its unyielding nature acts like a solid & the least resistance being through the vessel that must be torn to pieces at once. No column of water must intervene between the torpedo and the vessel for it then acts like a shield to ward off effects which like a pillow lifts her up on a big wave preventing damage.

II. Peter S. Michie

Biographical Introduction

Peter S. Michie was born in Brechin, Scotland, on March 24, 1839, the second son of William Michie and his wife Anne Smith Michie. His father was a journeyman watchmaker. Michie's parents, converts to the doctrines of Swedenborg, imbued the boys with not only morals and religious habits, but with the "doctrines which [they] ... espoused." This was the beginning of a "deep, unvarying and earnest faith" which characterized Peter's entire life.[1] Religious controversy was especially sharp in Presbyterian Scotland at this time, and in 1843 the family emigrated to Cincinnati, Ohio, when Anne's older brother had settled earlier. William established himself again as a watchmaker, and was eventually succeeded by his son, also named William.

The Michie family was "industrious and sober" and soon became outspoken abolitionists. The new Republican Party offered them a political avenue to express their beliefs. Peter, after completing the course at Woodward High School, where he won a gold medal for achievement, went to work at the Niles Machine Works as an apprentice, and "by his intelligence, industry and mechanical aptitude" soon became a foreman at the factory. He learned of a vacancy at the United States Military Academy at West Point, N.Y., for the district in which he was living and immediately supplied himself with letters of recommendation from his teachers and called at dawn at the residence of congressman George H. Pendleton before the legislator was out of bed. Undeterred, young Michie waited until Pendleton would see him. He presented his case "manfully" and won Mr. Pendleton's sympathy. Although Pendleton was a Democrat, and later ran for vice-president with George B. McClellan, he was deaf to all partisan claims on the coveted spot and sent "the young Republican mechanic his cadet warrant." Pendleton followed Michie's career with both interest and friendship until the politician's death in 1889.[2]

Michie's employers immediately offered him a salary higher than he might earn in his early years in the army, but he refused and immediately traveled to West Point for the preliminary examination. He easily passed and was enrolled on July 1, 1859. By the end of the following year the secession crisis had reach it peak with the withdrawal of South Carolina from the Union following the election of Abraham Lincoln. Michie ranked either first or second in every course he took at West Point. His main rival, and best friend, was John R. Meigs, son of Montgomery C. Meigs; the young Meigs would not survive the war. Indeed, all of John's correspondence was found among Michie's papers, at the time of his death decades later.[3]

Michie was well remembered, for not only was he a friend with Meigs, but with all of his classmates and the rest of the cadet corps. Due to the shortage of officer instructors, Michie was detailed as Acting Assistant Professor of Mathematics and taught the lower class. Although teaching took time away from his own studies, he managed to graduate second in his class (Meigs was first) on June 11, 1863. He was appointed to the Engineer Corps with the unusual rank of First Lieutenant.[4]

In June 1863, the war was at its height. Gen. Joseph Hooker had been defeated in Virginia; Gen. William Rosecrans, in Tennessee. With the exception of Gen. Ulysses Grant, all Federal aggressive operations were at a standstill. Upon graduation, Michie and the other graduates were sent on furlough, they thought for the usual two months. He had hardly received his commission and returned home when orders arrived for him to report for active duty. He had become engaged to Marie L. Roberts, a long-time sweetheart and graduate of the Woodward High School. His orders reached him on Saturday, June 20, 1863, and government offices were closed and the clerk of court was in the countryside. He wished to marry immediately before leaving for an indefinite period. He hired a horse and buggy, rode around until he found the clerk, brought him back to town and by midnight had secured a marriage license. He was married the following evening, had a hasty wedding dinner, and boarded a train for New York. From there he went to his first posting in Gen. Quincy A. Gillmore's Department of the South, then laying siege to Charleston, S.C.[5]

Arriving on June 29, 1863, and devoid of practical experience, he was immediately assigned to constructing batteries on the north end of Folly Island. Through August and September he worked day and night, "from the first superior, not only to the sense of danger, but also to that of fatigue." He then laid out the approaches to Battery Wagner on Morris Island. Subsequent to the fall of Battery Wagner, Michie was involved in the repairs to that battery as well as to Battery Gregg, which took until November 1863. He then was assigned to design and construct the Federal works on Cole's Island at the mouth of the Stono River. On January 16, 1864, he was appointed as Chief Engineer of the Northern District of the Department of the South.[6]

On February 6, 1864, he accompanied Gen. Truman Seymour's expedition into northeastern Florida as Chief Engineer. He participated in the battle of Olustee and directed the building of defenses at Jacksonville, Yellow Bluff, Palatka, and on the St. Johns River. Seymour characterized Michie as "always ready, always brave, always skillful." By April, the campaign had lost its aggressive aspects, and Michie was ordered to again report to Gen. Gillmore outside Charleston on April 13.[7]

General Gillmore, along with most of his force, was now moved to Fort Monroe as the X Corps, to be half of the Army of the James under the command of Gen. Benjamin F. Butler. Michie accompanied Gillmore and, on May 1, 1864, was appointed Assistant Engineer of the Army of the James. In that capacity he came to know well Generals Butler, William F. Smith, Godfrey Weitzel, John W. Turner, Adelbert Ames, and later George G. Meade, Andrew A. Humphreys, Winfield S. Hancock, and Ulysses Grant. Gillmore tried to retain Michie on his immediate staff by appointing him as an aide de camp with the rank of captain, but Gen. Weitzel, an engineer officer himself, did not want to see Michie's talent kept from the army. He appealed through Col. Cyrus Comstock to Gen. Grant, and had Michie's appointment as Assistant Engineer confirmed with the retained rank of captain.[8]

Michie took part in the skirmishing and fighting near Drury's Bluff on May 14–16, 1864, and was involved in constructing defensive works along the James River from May until September. On August 1, 1864, he was appointed Chief Engineer of the Army of the James as well as of the Department of Virginia and North Carolina. Michie oversaw the constructions of pontoon bridges across the James River at Deep Bottom and Varina. He participated in the assault on Fort Harrison on September 29, and directed the construction of the line of works north of the James River to the newly captured fort. He supervised the construction of the Dutch Gap canal from September 20 through December 2, 1864.[9]

Gen. Butler recommended for Michie a double brevet as Captain and Major which was effective from October 28, 1864, for "gallant and meritorious services during the campaign of 1864 against Richmond, Va."[10] Effective on January 1, 1865, Michie was brevetted as Brigadier General of Volunteers for "meritorious service in 1864."[11]

Michie directed all engineer operations of the column on the left of the Army of the Potomac during operations against Hatcher's Run and in pursuit of Lee's Army of Northern Virginia until the capitulation on April 9, 1865. Although Michie was Chief Engineer of the Army of the James, he served with the Second and Sixth Corps of the Army of the Potomac in the final pursuit of Lee. Michie organized two battalions of New York Volunteer Engineers and two companies of heavy artillery acting as pontoniers and engineer troops, together with their trains, into a flying column which outmarched the columns of the Army of the Potomac and were the first to reach the Appomattox River at Farmville, Virginia.

For these final services he received his brevet as Lieutenant Colonel, effective from April 9, 1865, for "gallant and meritorious services during the campaign terminating at Appomattox Court House, Va." He was given the rank of Captain in the Corps of Engineers on November 23, 1865.[12]

By the end of the war Michie had gained the confidence and respect of every general with whom he came in contact.[13] Indeed, in March 1865 Grant wrote that "his services eminently entitles him to substantial promotion and they will in the end not go unrewarded." He later wrote "He is one of the most deserving young officers in the Service."[14]

Following the Confederate surrender, Michie was stationed in Richmond from April 9, 1865, until April 30, 1866. During this time he studied the various campaigns in Virginia and their fieldworks, which he documented in a series of maps and a prepared report.[15] It was during 1865 that *Notes Explaining Rebel Torpedoes and Ordnance* was written and the illustrations prepared.

From May 1, 1866, to April 1867, Michie took a leave of absence, "preparatory to resignation," from the army. Benjamin Butler, on whose staff Michie had served, had acquired gold mines at Dysartsville in western North Carolina. Butler secured the services of Michie and sent him, as "Cashier and Superintendent" to direct his Mountain Mining Company. Michie was active in securing land for Butler, some of which was purchased by former Confederate General Robert F. Hoke. Although gold was found early, the initial optimism faded as productivity decreased, ore played out, supplies and food ran short, and expenses soared. By the end of 1866 Michie recommended selling the tract and requested to be relieved of his duties as of January 1867. Michie had by now clarified his thinking about retaining his commission in the Engineer Corps and returned to duty with the army in 1867 with the permanent rank of Captain.[16]

Professor Dennis H. Mahan at the United States Military Academy now sought out Michie, who was detailed there to teach. He was appointed Principal Assistant Professor of Engineering on April 20, 1867. He was named Instructor of Practical Military and Civil Engineering, Military Signals and Telegraphing, as well as Acting Assistant Professor of Chemistry, Mineralogy and Geology, on August 31, 1867. The latter assignment lasted only ten months. In 1870 he was dispatched to Europe as part of a group of distinguished engineers to examine the fabrication and use of iron in sea coast fortifications.[17]

On February 14, 1871, Michie was appointed Professor of Natural and Experimental Philosophy at West Point.[18] He served on the Board of Overseers of the Thayer School of Civil Engineering at Dartmouth College from 1871 to 1901. He was awarded a Doctor of Physics degree by Princeton University in 1871.[19]

Michie authored *The Elements of Wave Motion Relating to Sound and Light* (1882), *Ele-*

ments of Analytical Mechanics (1886–87), and *Elements of Hydro-Mechanics* (1888), and coauthored *Practical Astronomy* (1891).[20] In 1885 he wrote the biography of one of his best friends, *The Life of Major General Emory Upton.* During the final year of his life he completed a biography of Major General McClellan for the Great Commander Series, but he did not live to see it published.[21] Michie devoted his entire postwar career to teaching cadets at the Academy and was highly regarded by all who knew him, both faculty and students.[22]

Prof. Michie in the early 1870s (United States Military Academy)

Michie and his wife had two sons and one daughter. One son, Dennis, was killed in battle at San Juan Hill, near Santiago, Cuba, on July 1, 1898. Shortly thereafter the second son, William, a rising civil engineer, died of pneumonia. Michie had begun to show signs of cardiac failure during 1900 and was planning his retirement when he died of pneumonia on February 16, 1901. The Academic Board of the Academy passed a resolution declaring that Michie "exemplified in his official and private life those qualities of integrity, devotion to duty, and professional intelligence which this academy seeks to impress upon its graduates."[23]—H.M.S.

Editorial Notes

Few changes have been made to the text. A rare misspelled word has been silently corrected. The few apparent absent words have been inserted in brackets, *e.g.*, [is]. Where a word is incorrect, such as a verb tense, it is followed by [*sic*]. Where a word is clearly legible but not present in the unabridged dictionary, it is followed by [*sic*]. The rare word that was indecipherable is indicated as [illegible]. I have silently added commas for clarity.

I have included the manuscript page numbers within the text as **[5]**. These obviously differ from the page numbers of the book. A table of contents keyed to the manuscript pages precedes the book (there was no table of contents in the original manuscript).

The manuscript consists of eighty handwritten pages with a cover letter. It is associated with twenty-one watercolor plates. I have included the plates within the text where it seemed appropriate, and have also enlarged portions of the plates and inserted the enlargements as appropriate. Some of the details discussed, such as point lettering, *e.g.*, *a* or plane *cd*, may not be clear in the copies of the plates, and some are difficult to see even with full magnification, but this is not too great a hindrance in understanding the working of the various devices discussed in the manuscript. I have copied the plates on the web site listed below; all detail can be seen there.

A few drawings are present in the body of the manuscript itself; they have been included where appropriate. The manuscript document appears to be a copy, rather than the original report submitted to the Chief Engineer in Washington, D.C. Portions of the report, with small engraved copies of Michie's plates, was incorporated into W.R. King's *Torpedoes: Their Invention and Use*, which was published in 1866.[1] The original plates were returned to Michie at a later date.

Included with the colored plates is an engraved sheet including Cyrus Comstock's report on the assault on Fort Fisher where electric subterranean torpedoes were found. The report is accompanied by an engraving of the part of the ignition mechanism for the torpedoes.[2] Similar diagrams are included in the colored plates meant to accompany Michie's report and I have omitted a copy of the engraving in Comstock's report.

The plates were provided to me by the library of the United States Military Academy as TIFF copies on a DVD. For the illustrations in this transcription, I converted them to 300 ppi JPEG copies, enhanced the color, and tried to correct the white balance before cropping and sharpening. To view the Michie plates in color, with all the detail, go to the following web site: http://picasaweb.google.com/herbertmschiller/20091003MichiePlates IXXIForCD?authkey=GvlsRgCP368IiK450tigE&feat=email#.—H.M.S.

Contents

(manuscript page numbers)

Notes Explaining Rebel Torpedoes and Ordnance

Peter S. Michie

Brevet Brigadier General, United States Army
Chief Engineer, Department of Virginia and North Carolina

Office Chief Engineer
Dept. Virginia
Richmond, Va.
October 1865

Brig. Gen. Richard Delafield
Chief Engineer, U.S. Army
Washington, D.C.

General:

I have the honor to submit herewith notes explanatory of the rebel torpedoes and ordnance delineated on the accompanying sheets. This information has been obtained from the best sources within my reach.

I am indebted to Mr. Elliot Lacy[1] and Mr. John F. Alexander for the greater part of the information.

I am General

Very respectfully
Your obt servant

Peter S. Michie

Peter S. Michie
1st Lt., Bvt. Lt.Col. U.S. Engrs.
Bvt. Brig. Gen. U.S. Vols.

Notes Explaining Rebel Torpedoes and Ordnance As Shown in Plates Nos. 1 to 21 Inclusive

[5]

Plate No. 1
Spar Torpedo

This torpedo consists of a tank of sheet copper with brazed joints, somewhat egg-shaped as shown in Plate No. 1.

The essential feature of the exploding arrangement resides in a sensitive primer with a cylindro-conical head communicating with the magazine of the torpedo, and which is in contact with, and protected from the water by a thin hemispherical cap of soft, well-annealed copper.

This contact of the primer with covering cap, is effected by sliding up the small hollow cylinder which holds it, until it is pressed lightly against the cap, and fixing the cylinder in its position by turning in the setscrew shown on the side.

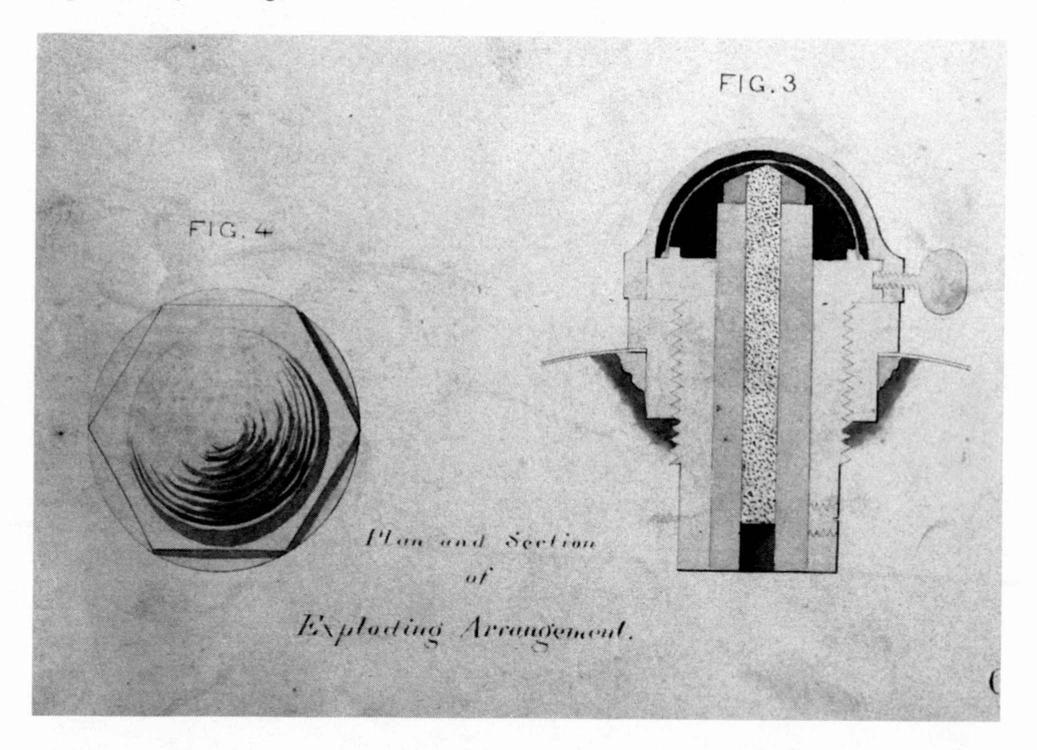

With this arrangement, the thin cap while completely excluding the water would [6] transmit any blow on it to the head of the primer, and exploding it communicate fire to the charge of the torpedo. This cap is carefully soldered to the outer cylinder, which contains the primer cylinder and forms the body of the fuse.

The safety cap which is placed over, but does not touch, the thin copper cap, and protects it from accident, is a brass cover sufficiently unyielding not to be indented by any blow that it would probably receive in service.

It was generally the practice to glue or tie a piece of shellac paper over the lower opening of the primer cylinder to protect the primer as much as possible from the effects of moisture.

Thus prepared, the whole is screwed into a bushing which has been previously braised firmly into the body of the torpedo, a little white lead being introduced [7] into the thread of the screw and a leather or gum collar placed under the head to render the joint water tight.

The torpedo is fixed at the extremity of a spar by means of a socket riveted on its smaller end. The other extremity of this spar, which is about 25 or 30 feet long, is attached to the boat carrying the torpedo by a goose-neck, which passes through a socket firmly bolted to its bow, thus permitting the torpedo to be unshipped and taken aboard, elevated out of the water, or lowered into it as the case may require.

The torpedo is maintained in its position in front of the boat and maneuvered by means of two guy ropes attached to the spar and passing thence to the sides of the boat, where they are connected with some tackle arrangement.

[8] When elevated out of the water, in which position it is carried when not about to strike an enemy, a third rope passing over the stem of the boat is also employed.

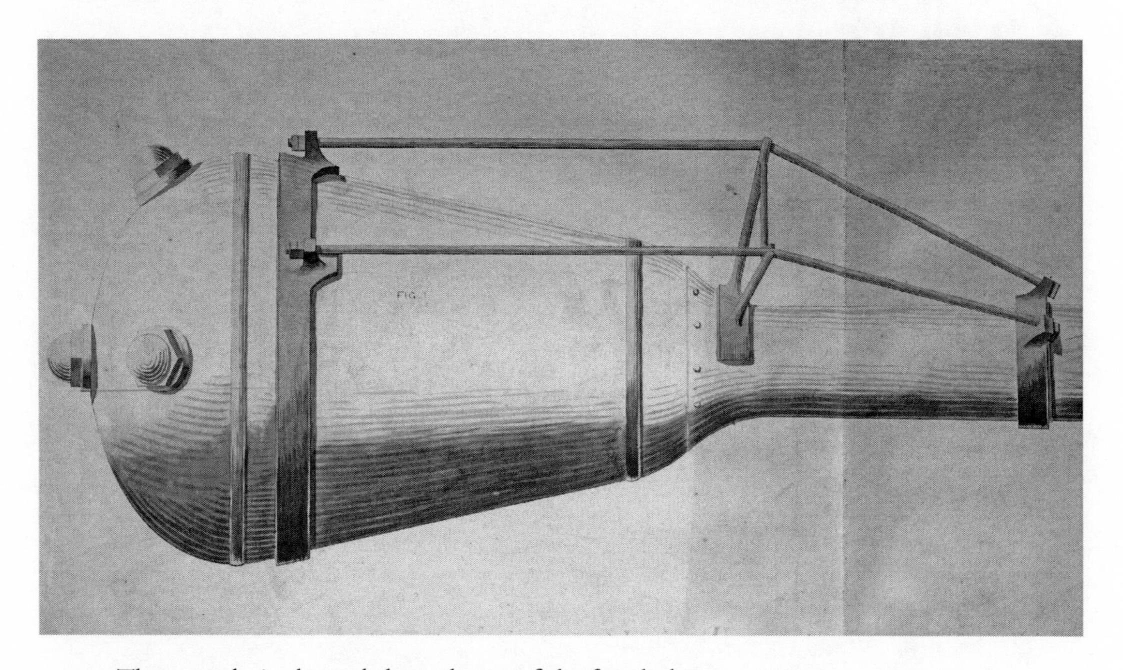

The torpedo is charged through one of the fuse holes.

The larger tanks were supported by braces of iron, as shown in the drawing, to prevent their weight from tearing them from their socket when raised from the water.

As long as the torpedo remains on board the safety caps are kept on to prevent accident.

When rigged for use it is kept suspended in the air until about to be used in order to prevent premature explosions by striking anything which might be floating or submerged in the water.

The sensitive primer used in the [9] exploding arrangement of these torpedoes was the invention, and was manufactured immediately under the eye of General [Gabriel J.] Rains, the chief of the Torpedo Bureau, who carefully preserved from publicity all information concerning their fabrication.

So sensitive was the detonating composition which was used, that a pressure of seven pounds applied to the head of one of the primers could explode it.

Chemical analysis would of course shed some light on its character.

The object in adopting an egg-shaped tank was to bring the bulk of the charge as near the objective struck as was compatible with facility of passage through the water.

When these tanks were first employed as torpedoes, they were mostly improvised out of soda-water fountains.

They were then supported [10] at the end of the spar by a basket arrangement of iron straps.

The primer was not the original device for exploding them, but the sulphuric acid arrangement shown in the accompanying sketch.

It consists of a small glass tube of sulphuric acid, contained in and resting against the head of the soft lead cap *K*.

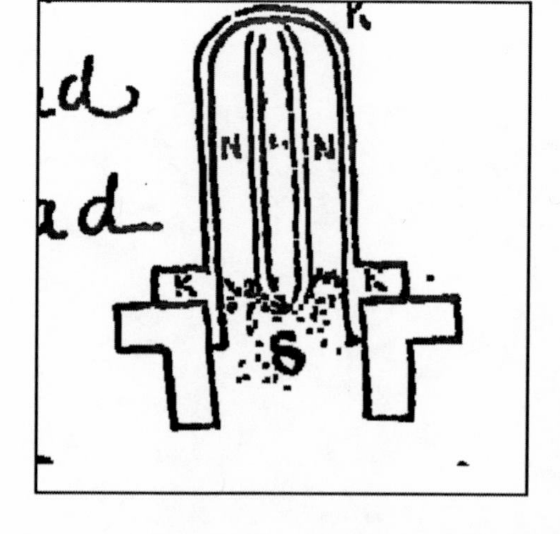

Surrounding the tube and holding it in position, is a mixture of chlorate of potassa and white sugar *N,N,N.*

S is a quick burning composition.

Upon striking any rigid body the soft lead cap is crushed, breaking the glass tube in contact with it, causing the sulphuric acid to come in contact with the chlorate thus producing fire, which is communicated to the charge by the priming *S.* [11] To secure a ready submergence upon slackening using the guy ropes, it was frequently necessary to load the torpedoes with weights, overcoming the disposition to float occasioned by the buoyance of the spar and the imperfect filling of the tank.

The spar was bent upward at its forward end to bring the torpedo into a horizontal position although inclined itself.

Every description of boat from a steam-tug to a canoe have at one time or other been armed with this torpedo. More recently, however, special boats were constructed, which were propelled by steam and capable of running at a very high speed. The after portion of these boats was covered with a bullet-proof sheet of boiler iron, for the protection of the [12] the [*sic*] crew and engine.

It might be anticipated that the explosion of the torpedo would occasion a frightful back thrust on the spar, but the effect in practice is too trifling to be noticed.

Plate No. II

The Swaying Boom Torpedo

The design of this torpedo was to obviate a defect which had been found to exist in all previous ones, that of facility of removal from their site.

The experience of causing them to explode when attempted to be removed had been adopted, but manifestly could not answer the desired end, as they were gotten out of the way and the streams cleared for vessels as [13] effectually by explosions as if by dredging as means existed for making the dredge perfectly safe to those performing it.

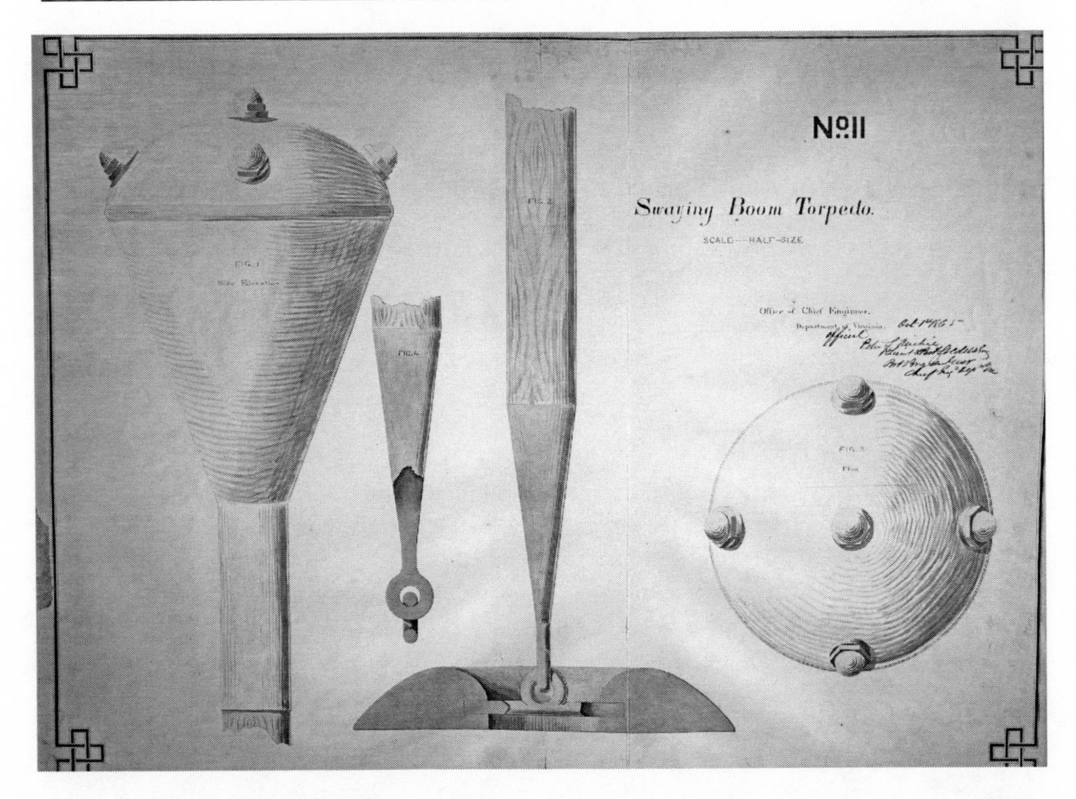

This engine is represented in Plate No. 2. It is composed of a copper tank of the shape and dimensions there shown, fitted with five exploding caps as depicted. These caps are the same in every respect as those already described as used in the Spar Torpedo, the object of this identity being to simplify as far as possible the material of the service.

This tank is fixed securely at the end of a boom, by means of a socket riveted to its bottom. The other extremity of this boom is connected by a double hook forming a universal joint with a flat annular anchor.

The charge of powder in the torpedo was such as to leave it a [14] buoyancy of between 10 and 15 pounds which has to be increased somewhat at first to allow for the decrease consequent upon the water logging of the boom.

The extent to which the arrangement described secured the desired point of submersibility will be best arrived at when we come to consider briefly the means usually employed for clearing a water course.

It was simply a chain, or a line sufficiently weighty to sink it, which was dragged perpendicularly to its length over the suspected locality. With all preexisting methods of anchoring, the torpedo would either stop the drag line, be exploded, or dragged out. In the two last cases the end of removal would have been accomplished, and in the first the precise location of the torpedo would have been ascertained and it could afterwards be grappled with [15] claw hooks and be drawn out.

On the contrary with the Swinging Boom Torpedo, no indication of its presence would be given when hung by the line, for owing to its slight buoyancy it would sink out of its way, allowing the line to pass over and resume its upright position when freed. In one of the preliminary experiments with one of these torpedoes a buoyancy of 20 pounds (nearly double the usual amount) was given it and a small leaded line used in dragging.

The distance and susceptibility with which a line drawn taut in the water transmits the vibrations caused by a jerk or blow are well known.

Notwithstanding all this, so easily did the torpedo sink and allow the rope to pass that it was utterly impossible [16] to tell when they came in contact.

To facilitate the passage of the line, the articulation which connects the boom with the anchor is placed below the level of the top of the latter as shown in drawing, and the surface of the boom and tank made as free from all inequalities as possible.

Plate No. III
Turtle Torpedo

To protect these torpedoes from whatever raking or grappling arrangement might be carried by the "devils" or other vessels searching for them, they were connected by a small line about 130 or 140 feet long with a second torpedo which was placed down the stream from them in the direction in which it was supposed any enemy would approach.

This combination is shown in Plate No. III fig. 1 [17] This second torpedo, which was

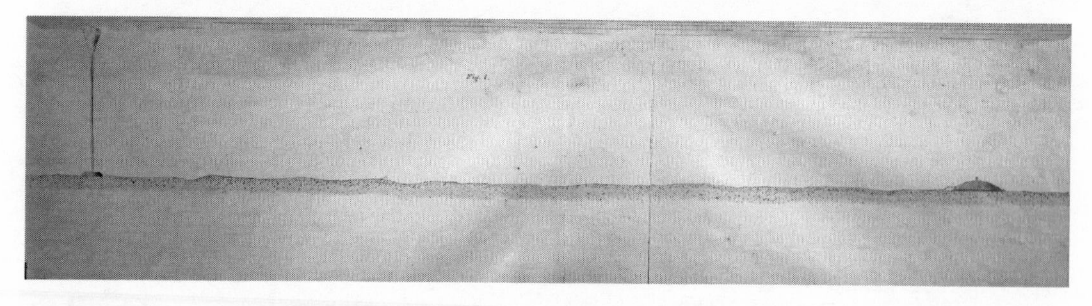

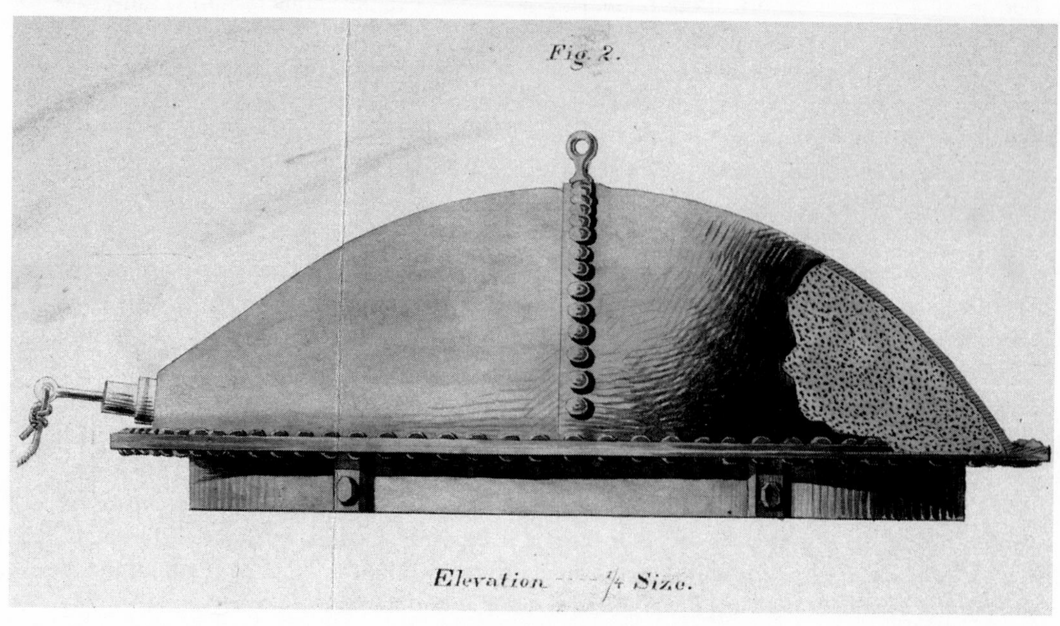

called a turtle from its peculiar shape was made of boiler iron and had a heavy plate of cast iron attached underneath to anchor it firmly. Its exploding arrangement which is shown in fig. 3 of the same plate was composed of a friction composition surrounding a roughened copper wire. This wire is attached to the lower end of a piston which has a packing of tallowed tow or cotton in an annular space around it, and a lead cap on the top of the fuse, to which it is soldered around its edge.

This allows the piston and with it the wire to be pulled out and the fuse fired, yet protects it from water.

To the eye at the upper end of the piston, the line from the Swing Boom Torpedo is fastened. The mutual actions of these two torpedoes thus is this, [18] if a line be used in dragging it will pass over both, which a "devil" would be unable to reach the turtle, but might hang the swinging boom. In this case, the outward movement of the vessel carrying it would take the torpedo along with it, thus tightening the line connecting with the turtle torpedo, drawing out the fuse piston and exploding it, in what would be from its advanced position just about under a vessel.

The whole of this arrangement was generally known as the "devil" catcher or circumventor.

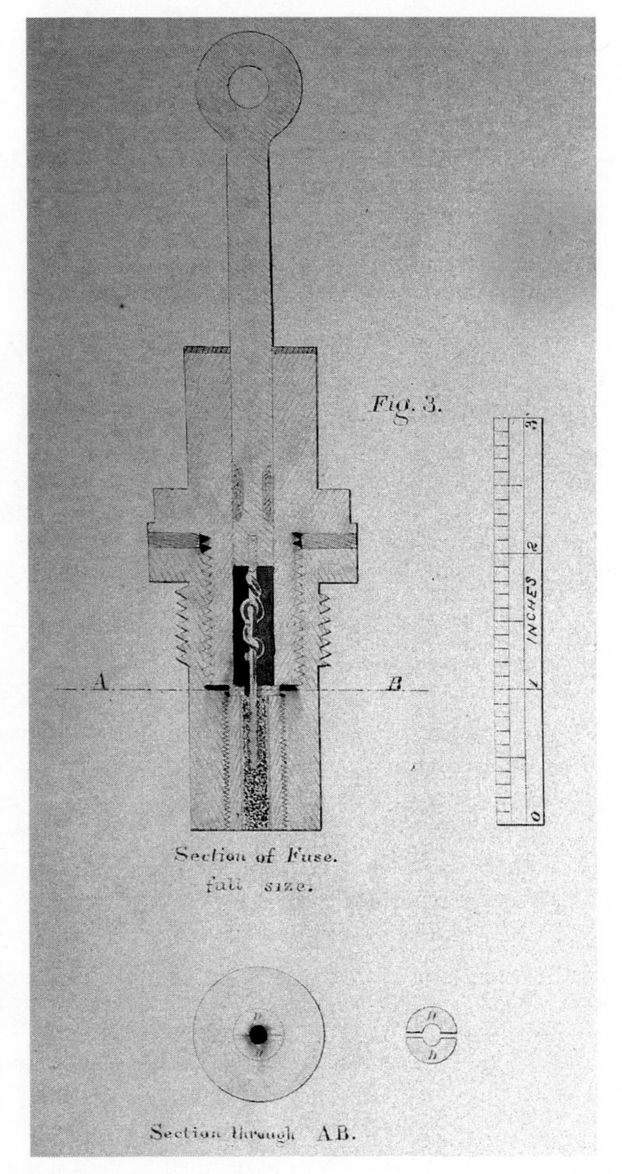

Plate IV

Blakely's Gas-Chambered Rifle Gun

This gun, which attracted much attention among artillerists in the South by reason of its large caliber, was the [19] invention of Captain [Alexander T.] Blakeley of the British service.

Two of them were brought into Wilmington, N.C., by the former privateer *Sumter* in the fall of the year 1863.

They were carried to Charleston, S.C., after some delay, where they were placed in position on what is known as the Battery. One of them had its base of breech cracked by improper loading but was subsequently restored in a measure and remained with the other in the city up to the time of its evacuation by the southern forces when both of them were destroyed.

The drawing shown in Plate No. IV was made from notes and a description given by an ordnance friend who had examined the gun.

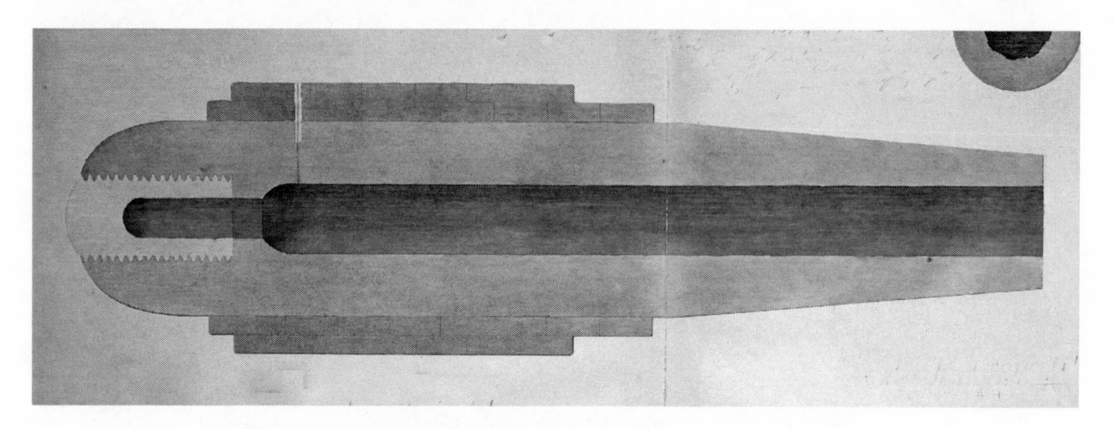

The essential features and dimensions of the piece will in the main be found **[20]** nearly correct, those in the interior are thought exactly so.

These guns were fabricated at the Mersey Iron and Steel Works, Liverpool, and were composed of a cast iron body reinforced after Professor Headwell's plan with bands of wrought iron shrunk on in position.

From all that could be ascertained it seemed that the trunions were not cast on the shell of the gun, but forged on one of the reinforce bands which was made heavier than the rest for this purpose.

The chamber had been formed by counter boring from the rear through the base of the breech and screwing in a hollow bronze plug as shown in the drawing.

The essential feature of this gun resides in the manner in **[21]** which the chamber was to be used.

The cartridge instead of being formed as usual to fit the chamber of the piece was merely placed against the mouth of it without entering, the projectile resting well home on the charge. Thus the chamber would constitute an air space in rear of the powder and the projectile [*sic*] which would be filled by gasses produced by the combustion of the charge and hence the name gas-chamber in contradistinction to powder-chamber.

The action claimed for this chamber was a great reduction of the strain on the gun without a proportional diminution of initial velocity. The result being effected in somewhat this manner.

It is well established that the most powerful and bursting strains which a gun endures, is in the incalculably **[22]** short space of time between the ignition of the charge and the first movement of the shot. In other words the difficulty is in overcoming the inertia of the projectile.

In a gun furnished with a gas chamber this strain was reduced by allowing space for the gas first produced by the combustion of the charge to expand freely, the strain thus coming on more gradually. This point was based on the principle that these first gasses originating at the interior of the orifice of the vent as a center would in a state of high tensions be projected in all directions with great velocity. Those volumes thrown forward would expend their forces in overcoming the inertia of the shot while those in the other direction would be relieved by expansion into the gas-chamber.

The well known powerful action of a petard on a solid resisting body was adducted in illustration of the mode of [23] action.

That the forces developed in the burning of the charge operated somewhat in the manner set forth is clear, but whether this mode of operation would produce the effect claimed must only be decided by experiment.

Experiments which were made by Col. [George W.] Rains[2] of the Augusta, Ga., Arsenal by means of the electric ballistic pendulum and Rodman's indenting apparatus on a Columbiad in which a gas-chamber had been bored, showed a very decided decrease of strain, which was not accompanied by a correspondingly diminished initial velocity.

Experiments made at Charleston on the original gun lead to the same conclusion.

The results obtained by Major [Thomas J.] Rodman in one of his experiments is such as to shed some light directly on the point involved in the gun. [24]

When powder was burned in a space occupied by itself it gave a pressure four times as great as when burned in double its own space.

Now the charge of powder used in this gun was said to be equal in volume to the gas-chamber; this being so, the pressure on the gun would be reduced to one fourth of what it would have been without the chamber.

It is not thought that a projectile remains motionless in a gun until the pressure, on its base from the burning of the charge, becomes equal to its weight, but that the volumes of dense, heavy, and compressed gas which are hurled against it, put it in a state of motion at an earlier instant.

At Augusta, Ga., was manufactured for this gun a powder the grains of which were an inch and a quarter in diameter.

[25] The projectiles which came with this gun from England belong to the so-called button variety, the rifling being accomplished by means of copper flanges secured to the side of the shot, as shown in the drawing.

The mode of attachment is by letting into a dove-tail slot which was placed in the shot after turning, the flanges themselves being turned with the shot.

The twist of the grooves as ascertained were about one turn in sixty-five or seventy-five feet and was uniform.

Maury's Elongated Projectile for Smooth Guns

This invention which is shown in Plate No. IV, as brought [26] from England by a man named M.F. Maury (not formerly of the U.S. Navy) and proposed to the rebel ordnance department in March, 1864.

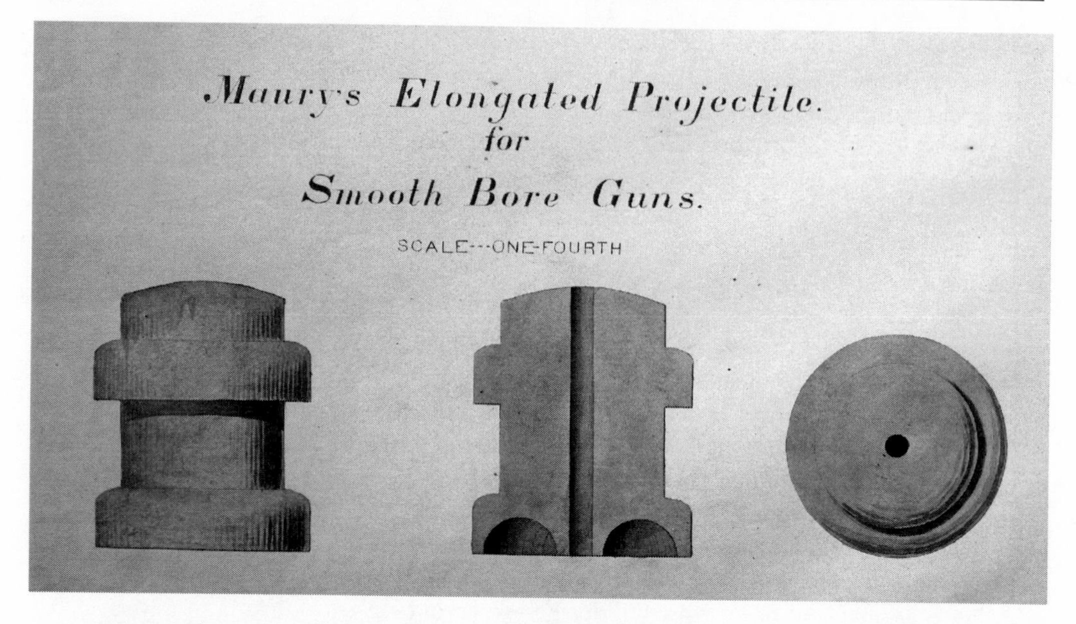

Maury's Elongated Projectile.
for
Smooth Bore Guns.
SCALE---ONE-FOURTH

Its peculiarity consisted in the central and axial aperture by which the alignment, while it was describing its trajectory, was to be maintained.

The annular cup shown in the base was for the purpose of bringing the center of gravity more nearly into coincidence with the center of figure, there facilitating the maintenance of alignment.

The Chief of Ordnance was impressed by it, for some of them were ordered to be cast for the purpose of experiment which was subsequently performed at Drury's Bluff.

Eighteen shots were fired in all, fifteen at long range and three at short range of about 300 yards into a [27] perpendicular bank, with results which from the minutes made at the time of the firing may be summed up as follows.

In all the shots at long range the absence of any other drift or erratic deflection from the line of fire than would have been experienced under like circumstance with spherical shot showed that a true alignment had been attained.

The shots fired at short ranges cut circular holes in the bank every time, which showed that the axis of the shot was tangent to the trajectory.

The gun used was an 8-inch Columbiad of the ordinary model.

The shot had a windage of only about two hundredths of an inch, the center aperture having a cross section of three-fourths of a square inch, and the weight of shot was 96 pounds. Ten pounds of powder was the charge used.

[28] A singular fact was observed in connection with shots fired at same elevation, which was that their ranges were about the same as would have been given by round shot a third lighter with the same charge and elevation.

The recoil of the piece was quite violent and the experiment was terminated by its being thus dismounted from its carriage.

Encouraged by these results, more shots were prepared and a subsequent experiment made at Chaffin's Bluff, which lead to identical conclusions.

When preparing for the second experiment it was proposed to the Inspector of Ordnance in the Richmond Arsenal, under whose orders the shot were prepared, that the central opening should be omitted in casting, or plugged as a means of ascertaining its precise effect. Whether the suggestion was acted [29] on is not known.

During the latter portion of the past year quite a large number of these projectiles were manufactured for use in the service, but whether any of them were ever actually employed in action is not known.

They were cast in chills for the double purpose of hardening and dispensing with the necessity of turning to gauge, which would have been required in sand moulds.

Toward the last they were modified somewhat from the original form by omitting the annular recess on the side.

The shot then presented the form shown in the appended sketch.

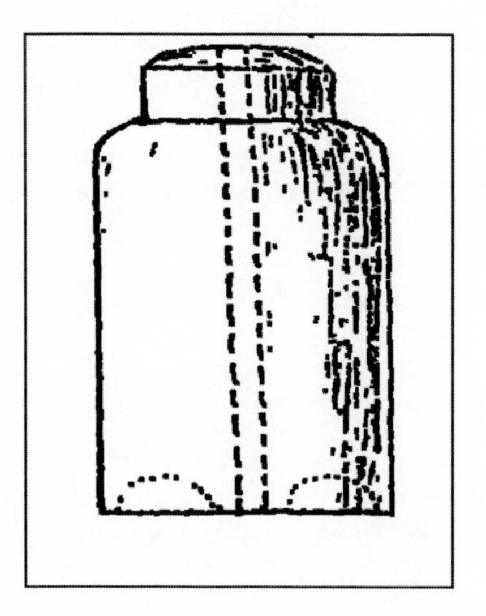

[30]

Plate V

Floating Tin Can Torpedo

This consisted simply of a tin can as shown in the drawing partly filled with powder leaving an air chamber above to give flotation.

The only case known of their use was on the 20th May 1864 when three of them were found under a schooner in the river along Jones' Landing. The simple fact of their being

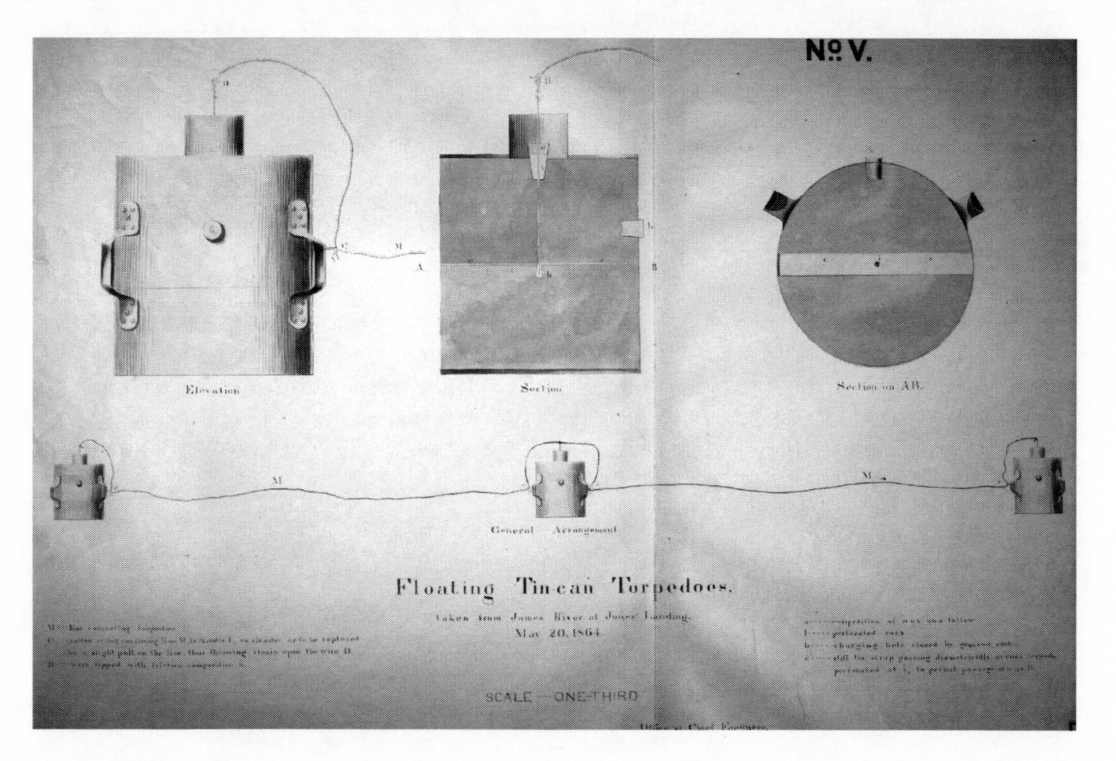

torpedoes frightened the men from the boat, but they were afterwards drawn ashore and examined. It was presumed that they were to be attached to a line a few feet below the surface of the water and exploded by the pull on the line as the vessel moved on, after breaking the cotton string binding the rope to the handle. This arrangement is of little value being more liable to failure than success.

[31]

Plate VI

Submarine Torpedo

This torpedo or submarine mine consists of cannon powder placed in a water-tight tank into the center of which pass the terminals of two insulated copper wires. Between their terminals is a bit of quill through which passes a fine platinum wire by which they are joined.

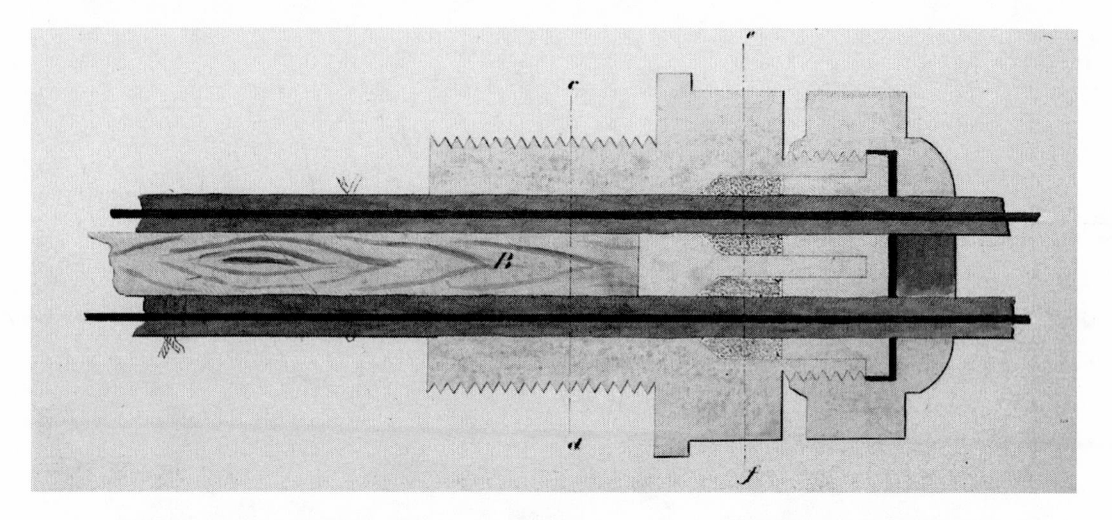

The bit of quill contains fulminate of mercury, and is surrounded by a small cartridge of rifle powder.

As this fulminate detonates at a very low temperature, it requires only a slight elevation of temperature in the platinum wire, which being effected by means of a galvanic current, explodes instantly [&] evenly the whole charge of the tank.

The tank is made of ⅝ boiler iron cut and riveted together in the shape of a spindle. The center piece [32] is cylindric having riveted into each end the bases of two conic frustum pieces, which receive into their tops two smaller, but similar shaped pieces of brass into which are screwed the lifting screw *C*, and the face plate *D*.

The nozzle is composed of three brass pieces. The cap-piece *g*, the packer *h, h', h"*, and the double-male screw *h,h"j,j"*. The cap piece is the ordinary female screwcap or a water plug with the exception of having an open face.

The packer is a cylindric disk, a quarter of an inch thick, into which are made two round holes, which have riveted into them pieces of brass pipe about 8/10 of an inch in length, and a tenth of an inch in thickness. The size of the holes [33] in disk as well as the diameter of its pieces of pipe vary to suit the insulated wire used.

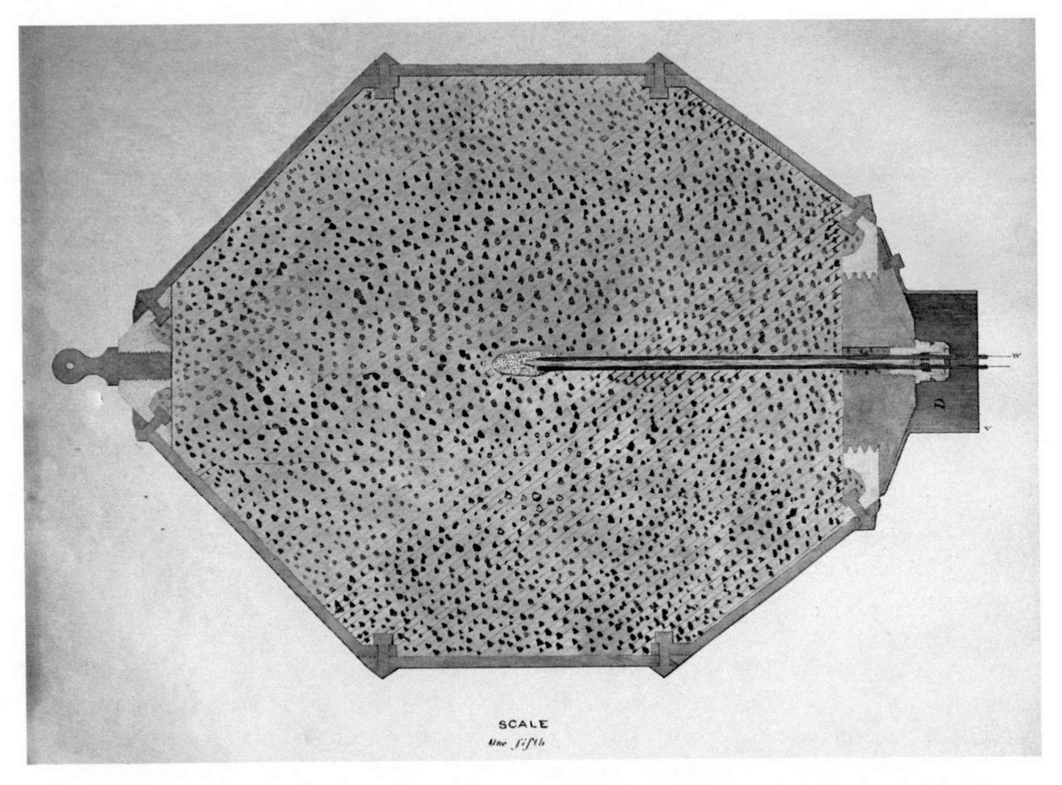

The double male screw is the ordinary double male hexagonal wrench screw with a circular flange. The bottom screw is hollow to circular flange *ll*, and top screw terminating in the hexagonal wrenchings has bored into it two cylindro-conic chambers *KK*, which receive the tubes of the packer.

The connections consist of a square strip of wood *B*, twenty-two inches long, one end of which enters the bore of the bottom screw, and is there wedged by four small pieces of wood as shown by the sections *cd*. The other end is shaved off to the length of the quill, *w*, which is fastened into a notch by waxed threads passing through it, and then passed through a hole in the wooden strip about a quarter of [34] an inch below this end and there tied. At right angles to this tying of the quill, waxed threads are passed over the center of the quill and again through the wood as shown by section *ab*.

The quill secured, two insulated wires are passed through the packer *hhh*, the cylindro-chambers, *KK*, of the double male screw, *hhjj*, and along the square wooden strip *B*, to which at slight intervals, they are securely tied with short lengths of marline, *m,m,m,m*. The terminals of these wires opposite the ends of the quills, are bent into this form , as the fine platinum wire *x*, passing through the quill winds its ends around them between these short bent curves, which are compressed so as to fasten the platinum wire.

One end of the quill is there stopped with a wafer of beeswax, the quill [35] filled with fulminate of mercury and the other end closed in the same manner. Around this end, a small cartridge of rifle powder is tied, as shown at *c*, in the tank.

The chambers *KK* being filled by stuffing tallowed cotton wick around the insulated wires as shown by section on *ef*. The packer is forced home by screwing on the cap-screw *g*.

The nozzle thus prepared, is now screwed into the iron-face piece *D*, and the guard *v*, made of a quarter of an inch sheet-iron, is put on by three short screws, one of which is shown in the section.

The tank is filled by turning it up on its guard, removing the lifting screw *C*, and pouring in powder through a copper funnel whose neck is curved to prevent the falling grains from striking the rifle cartridge.

[36]

Plate VII
Electric Subterranean Torpedoes

A, the cylindric tank which is made of quarter of an inch sheet iron riveted together, is 22 inches in length, 13½ inches in diameter, and holds 100 pounds of powder. *B*, is simply a 20 inch shell of 2 inches thickness of metal.

C, the conic tank is an old buoy altered so as to receive the nozzle.

The nozzle, excepting its size, is the same as that used in the submarine torpedo.

The connections are the same as those used for the submarine torpedo with this exception; instead of using the quill and platinum wire, a prepared fuse is connected to those terminals [37] which are intended for the interiors of the tanks.

This fuse consists of a cylindric wooden stopper rounded at its ends, 1¼ inches long and ⅝ of an inch in diameter, having a hole ⅐ of an inch in diameter passing lengthwise through its center, a tin tube 1¾ inches in length and ½ inches in diameter, and a gutta percha (2 inches long and ⅛ inch in diameter) cylinder, through which passes parallel to the direction of its axis two small copper wires.

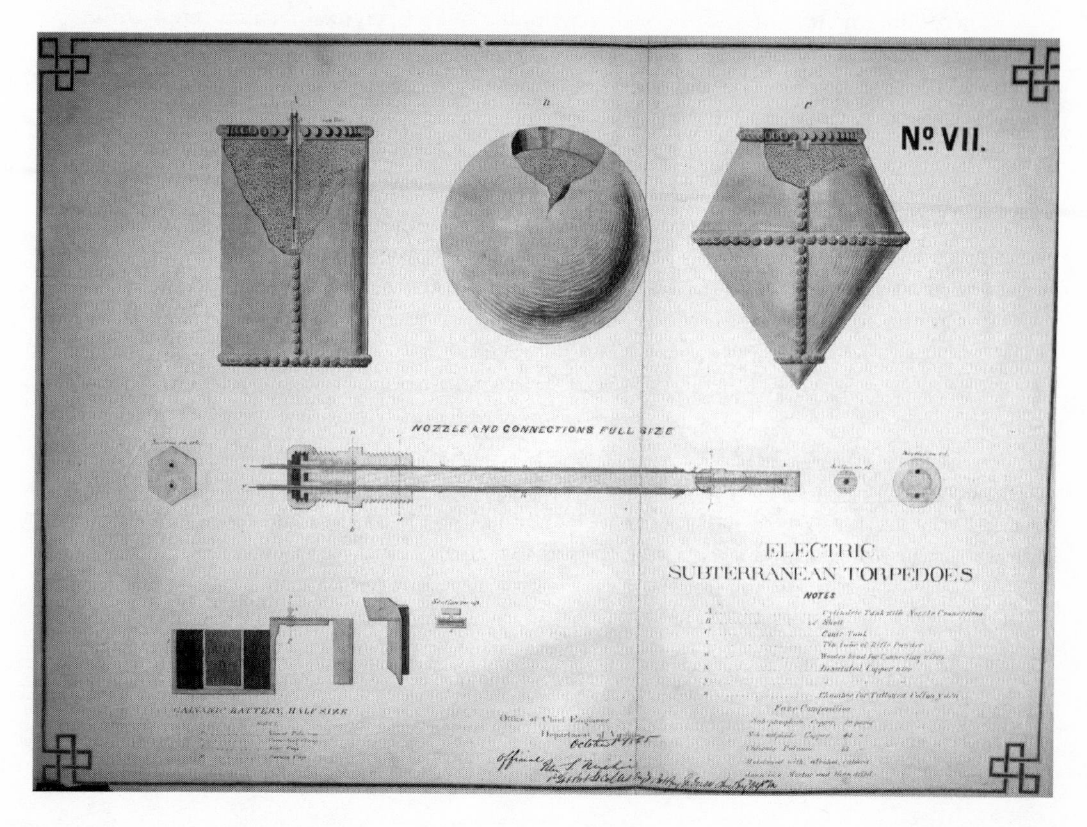

The stopper is turned off so as to give an entrance piece W, a half an inch long for the tin tube, which is punched in on the stopper to prevent slipping.

The two parallel wires shown by the continuous black lines diverge in passing out the exposed stopper end and pass along small grooves made to receive them **[38]** until they connect with two parallel copper eyelets *x'x',* which pass through the stopper head at right angles to the direction of the gutta percha cylinder as shown by section on *ef.* The other ends project from the gutta percha 1⁄20 of an inch, and are separated by about 1⁄50 of an inch, and are surrounded by a composition of 10 parts of sub-phosphide of copper, 45 parts of sub-sulphide of copper, and 15 [*sic*][3] parts of chlorate of potassa, previously prepared by moistening with alcohol, triturating in a mortar, and then drying; pack well up against them in a wrapping of thin sheet lead.

The tin tube is then filled with rifle powder, and its filling end closed with a thin layer of moistened calcined gypsum which soon hardens.

This fuse is attached to the wires which pass along the strip of wood **[39]** from the nozzle by simply notching the end of this strip to receive the stopper head and passing the wires through the eyelets in which they are wedged with small copper nails so as to insure perfect contact.

The galvanic battery differs from the "Grove's" in making the zinc cup so as to hold the diluted sulphuric acid and in using iron in place of platinum.

Plate No. VIII

Represents the connections for submarine and subterranean electric torpedoes

Submarine connections

The tanks being ready, they are each connected to two insulated wires previously stopped on to 3⁄4 manila or tarred rope at intervals of 20 feet **[40]** and of sufficient length to reach the shore from the point at which the tanks are submerged.

After submergence and anchoring against tide and current, while paying out the rope and its wires from the submerged tank to the shore they are anchored at intervals of 300 feet by lashing them to ketledges, so as to bring the rope and its wires on the bed of the river. On reaching the shore the wires are connected thus; one wire from each tank is numbered and laid into a prepared trench extending from extreme low water mark to the bombproof, or sheltering stand, a sketch of which is seen on section *AB*.

The other wires are cut off above extreme high water mark, their ends freed from insulating material, scraped and then connected to any one of **[41]** those wires lying in the trench as a center wire, by removing its insulation and wrapping around it these scraped wire ends. This being done they are reinsulated as shown in sketch connections at *G*.

The wires are then entrenched up to the bombproof and are carried, excepting that which is used as a center wire, to its berm where they are connected to the key plate in the following manner, Nos. 1, 2, 3, 4, or whatever arbitrary numbers used to discriminate among the submerged tanks, must be connected to the same numbers on the thin copper slips employed as circuit keys. See Key Plate and Battery connections.

The center wire is connected inside the bombproof to one pole of the Galvanic Battery, while from its other pole a wire passes to the berm and is there attached to the continuous

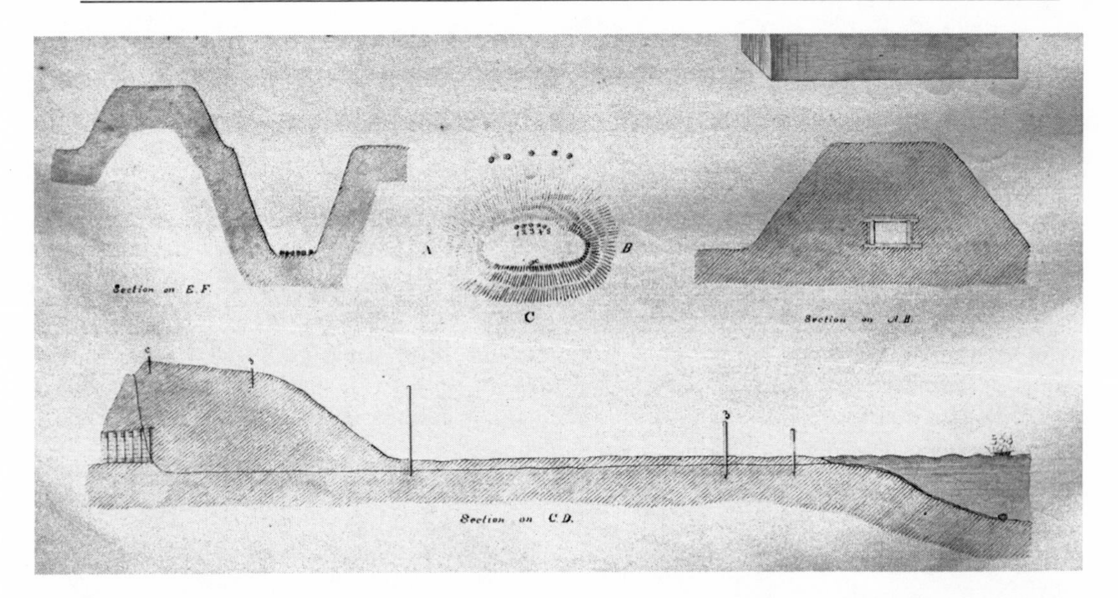

strip of copper lying immediately under the points of the keys [42] upon the key plate.

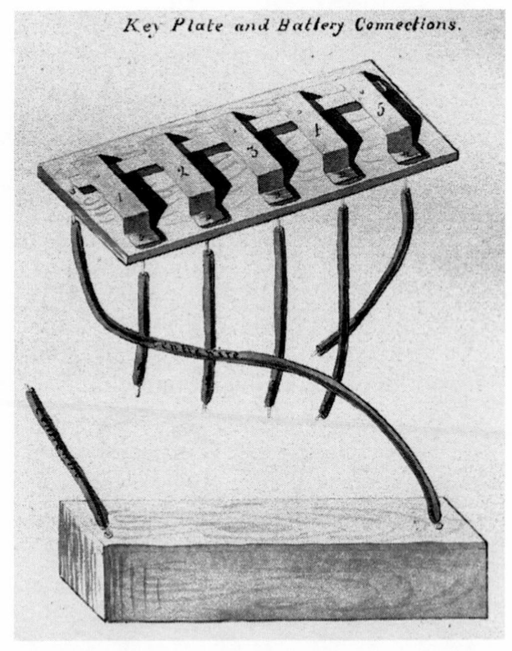

The galvanic battery is only charged when an enemy is approaching and then the center wire connecting the key plate is disconnected and removed from the key plate to avoid any accidental discharge of the tanks. As soon, however, as an enemy is within one or two hundred yards of the position of the submerged tanks the central wire is reconnected and the arrangement complete as an element of destruction in defensive warfare.

To explode for example No. 3 tank, touch the copper key marked No. 3 so as to bring its point in contact with the continuous strip of copper lying beneath it, and so for the remaining tanks. That the operator who is generally on the berm or lookout station may know [43] whenever an enemy is over a tank, stakes, called range stakes, are driven in the mound or parapet at follows: a center stake *C* is driven permanently into the superior slope allowing 6 inches to rise above the surface. From the crown of this stake at the time of location, sight to the point of submergence of each tank and drive in the direction of this line of sight stakes corresponding in number to the number of the tank, carefully observing that these little stakes have their tops exactly in line of sight from the top of the central stake *C* to those points on the surface of the water under which the tanks lie as shown by section *c,d.*

The tests usually employed in detecting extraordinary escape in the galvanic current to these tanks was that familiar [44] to telegraphic operators and known by them as the "tongue test." It was deemed sufficiently accurate for ordinary purposes since the length of cable employed seldom overreached a mile. If an escape exists between the tank and the battery, this was readily detected by the electrician whose duties were to attend to this and always insure a ready discharge of the tank. To be able to make the detection it was necessary for the electrician to test thoroughly the cable or wires before attaching them to the tank and to test immediately after laying the tank in its position in the sea, bay, or river. If for example tank No. 3 was to be tested, put one pole of the battery in connection with the ground and remove the center wire leaving its end suspended in the atmosphere, now bring the points of No. 3 wire and the wire from the other end of the battery [45] to the end of the tongue; if a sharp, acrid sensation with free secretions of saliva are experienced, you may be sure an escape exists somewhere between yourself and [the] tank, which if very heavy would make it necessary to overhaul the wire and find the defective point, in order to secure the discharge of the tank.

These torpedoes were used in defense of the inlets at Fort Fisher and Casewll at the mouth of the Cape Fear River, and it was one of this class of torpedoes which destroyed the steamer *Commodore Jones* in the James River in May 1864. These torpedoes have been known to explode twelve months after their submergence.

2 — Subterranean Connections

The subterranean torpedoes had for each group an entrenched wire one end of which joined a brass connecting [46] screw on the magneto machine or exploder and the other one of the two fuse wires which project from the nozzles of each tank constituting the group while the remaining projecting wire from the nozzle of each tank was either fastened to its exterior surface, as shown in the buoys and cylindrical tanks of group 3, or else joined to a plate of iron buried with the tanks as shown by *G'* group 1 and 5. The location of groups was by means of range stakes as in the submarine torpedoes, or by a small red flag placed in the center of each group.

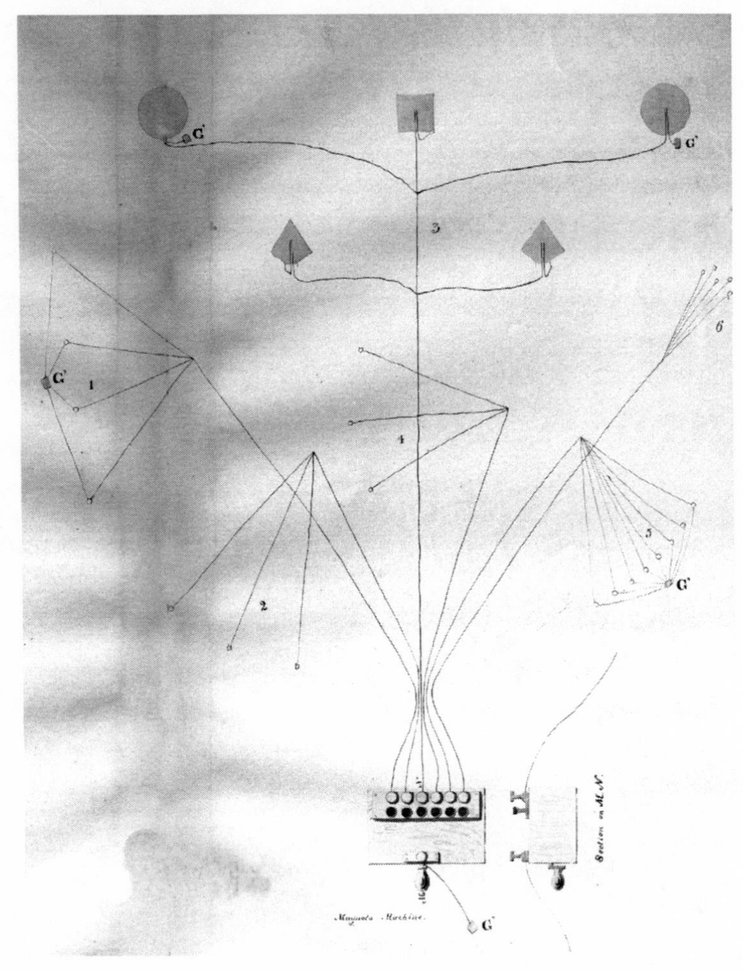

The groups were generally fired singly but

all could be fired at once. The *G'* shows that a ground plate is employed in the group whenever it occurs as seen in 1, 3 and 5, while 2, 4 and 6 use the exterior surfaces of the tanks and [47] plates as ground plates.

The machine for firing the groups was brought from England where it was known as "Wheatstone's Magnetic Exploder for Mines" from Prof. [Charles] Wheatstone, the inventor. The magneto machine in principle was simply an induction magnetic machine differing from those formerly used for medicinal purposes in having its armature to revolve instead of its helices and in being more powerful by reason of combining more horse shoe magnets and a greater number of helices. The connection plate is shown in the following sectional drawing.

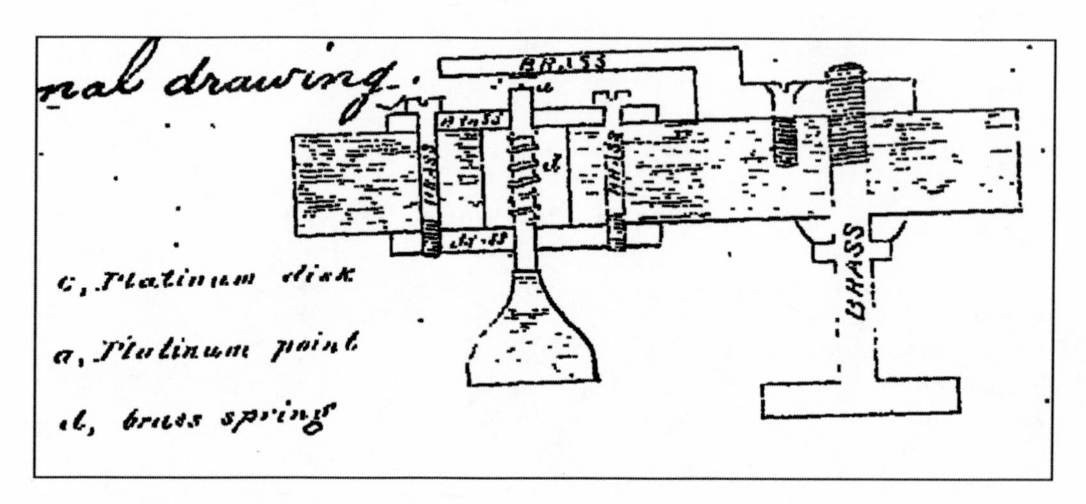

[48] The firing was by means of an induced current which starting from the machine on turning its handle fired the composition in its leap between the terminals of the small copper wires in the fuse.

Plate No. IX

Wrought-iron Concussion Shell

This shell depends for explosion on the heat which would be developed by impact on a solid resisting body like the iron armor or turret of a vessel.

The idea of thus exploding a hollow projectile is of British origin but the shell based upon it which is shown in figure 7, Plate IX is due to John M. Brooke, formerly of the U.S. Navy. [49] Practice with it on an iron plated target gave very satisfactory results both as regards certainty of explosion on striking and demolishing effect on target. To increase its destructiveness the head of the projectile was made heavier thus securing deeper penetration.

Wrought-iron Rifle Projectile

This missile which is shown in Plate IX figures 2 to 5 inclusive is another of Brooke's devices. Experiment had shown that cast iron bolts flew to pieces so easily on an iron shield as to be a serious defect where great battery was required and recourse was had to wrought iron which gave far better results.

The first shot made of this [**50**] material was built up by welding around a central core piece longitudinal slabs of a trapezoidal cross section.

The shot so forged were found to split open along the surfaces of the welds and [was] soon abandoned for the method of construction here shown which consists in welding wide collars around the core piece thus increasing the circular strength of the projectile.

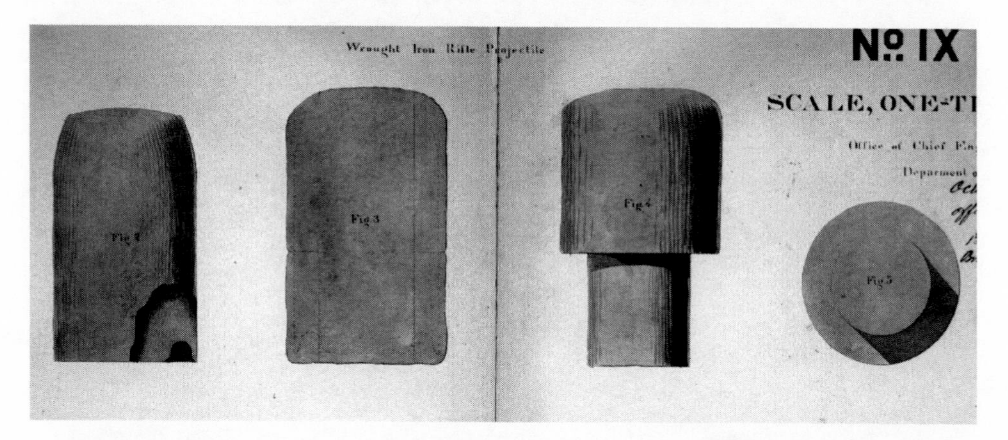

When so formed they were free from the defect of splitting and merely thickened on the head upon striking.

At some practice with a 7-inch Brooke Rifle and these bolts weighing about 125 pounds on a target 100 yards from the gun covered with three courses of iron [**51**] two inches thick by ten wide, the first shot broke through the first two courses and injured the third to some extent. A second shot in the same place went through the remaining course, through a 12 inch oak backing and some distance into the earth behind. This was the projectile which was used by the Brooke guns in the defense of Charleston, S.C., during the attack of the Union fleet in June, 1863, and more lately by the rebel ironclads in the James River.

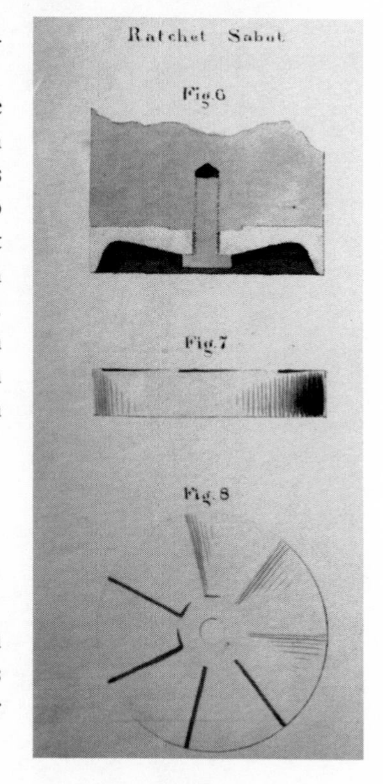

Ratchet Sabot

This invention, which is also one of Brooke's, is shown in Plate No. IX, figure 6. After extended experiments it was decided upon as being the best and used on all the heavier [**52**] cast iron rifle shot and shell.

Tennessee Sabot

This sabot, shown in Plate No. IX figure 9, preceded the Brooke Ratchet but as it gave an injurious strain on the gun was abandoned for the latter. The explosion of the charge flattened the concave saucer on the base of the shot and expanding it laterally forced it into the grooves of the gun with an almost unfailing certainty.

Ring Sabot

Shown in Plate No. IX figure 12 was a modification of the sabot used in the U.S. service for the purpose of economizing the use of copper. The rifling of the projectile was perfectly accomplished but it had the very objectionable defect of flying to pieces on leaving the muzzle of the gun.

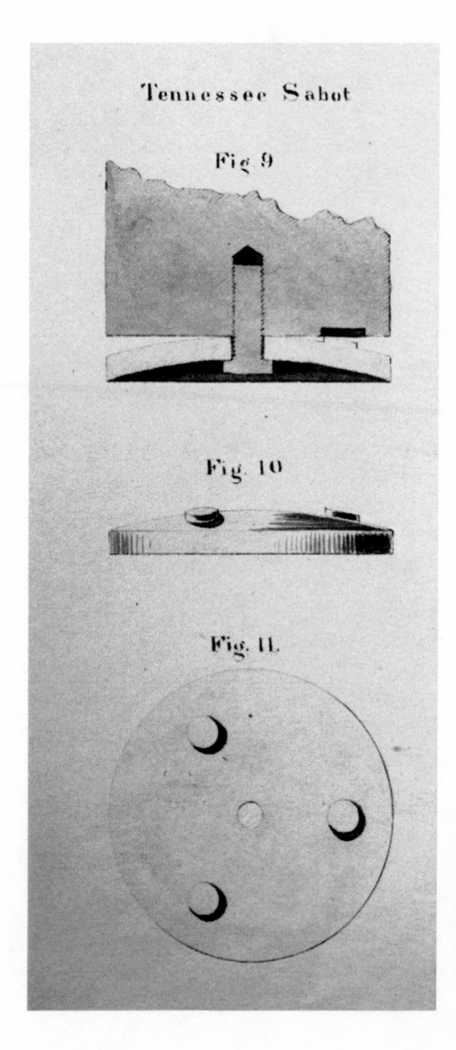

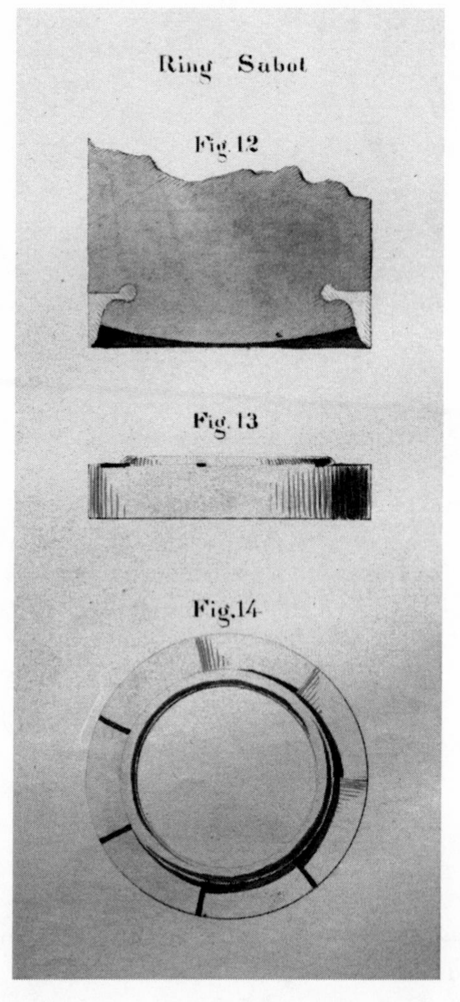

[53] Wooden Sabot with Culot

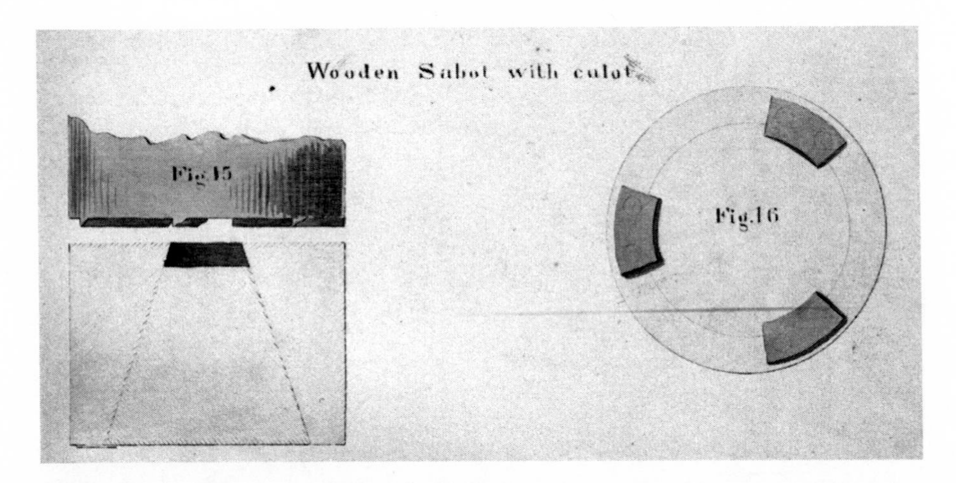

This shown in Plate IX fig. 15 was another copper saving device in which the rifling was accomplished by the expanding of a conical culot. The sabot imparted the rifle motion to the shot by impressing itself on the radial ratchets shown on the base. The three thin pieces of iron seen on the bottom of the sabot are for the purpose of retaining the culot in position.

Papier-mache' Sabot

This was for the purpose of accomplishing the same end in field artillery as was attained in heavy [artillery], by the wooden sabot. The attachment of the cartridge to the sabot was to expedite the loading and render it less liable to be broken in transportation.

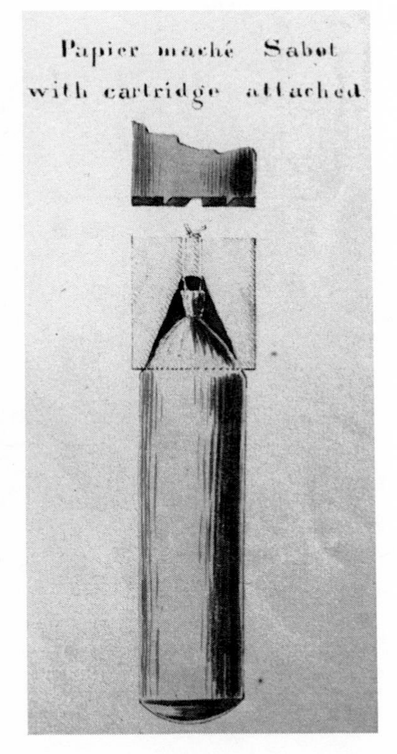

[54] Wrought-iron Spheroidal Shot

This projectile was devised by Brooke for use in his heavy banded ten and eleven inch guns at short range where great momentum and demolishing effect rather than accuracy were desirable. The rebel ironclads were supplied with these projectiles on the occasion of their raid down the James River in January 1865 and some of them were used against our ironclads on that occasion.

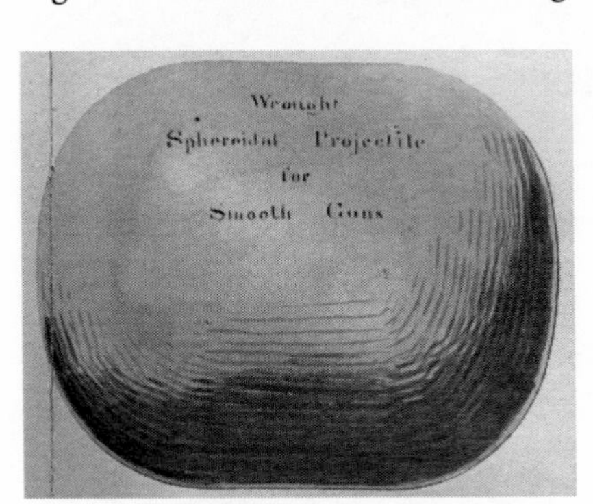

[Plate X]

River Torpedo

This torpedo which is shown in Plate No. 10 was the invention of a man named [Zere] McDaniel and being one of the earliest and most primitive of its class, is to be regarded as an illustration of this peculiar service when in its infancy and rather of historic value than of any inherent worth.

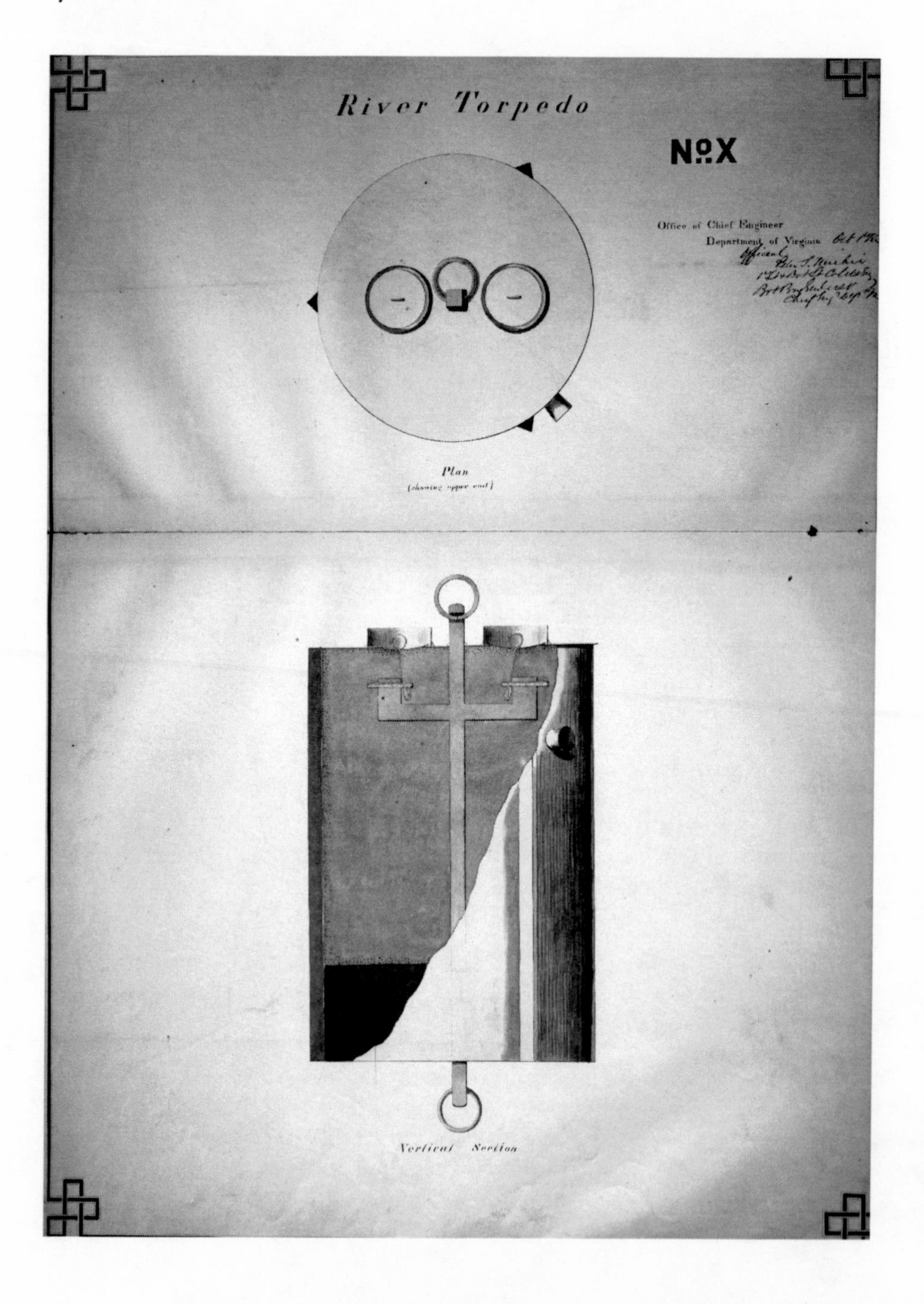

[55] The tank, when in the water, rests on its side being secured in this position by a short piece of rope whose ends were made fast to the eyes formed at the extremities of the iron stem which passes through the torpedo. This rope which passes beneath the torpedo has at its middle an eye to which is connected a rope that reaches down to the anchor. This torpedo is fired from the shore by means of a line connected to the looped ends of the wires of the friction primers. The same shaped disc of soft copper through which the primer wires pass allow them to be torn out of the composition by a pull on the line and yet occlude water.

Plate XI
Fretwell's Percussion Torpedo

[56] This apparatus is illustrated in elevation section and general arrangement in Plate No. XI. As far as ascertained this torpedo has done more execution than any other during the war. Its form, dimensions, method of anchoring &c. will easily [illegible] understood from the drawing.

It will be seen that the inverted saucer shaped plate *a* merely rests unfastened on the top of the torpedo the contact being around the lower edge of the central aperture in the plate. Into this aperture rises a low rim of sheet tin *d* which catching against the rectangular edge of the opening prevents the plate from sliding off when the torpedo is tilted by action of the current or tide.

Attached to this plate is a wire *b* branching into three parts which connect with as many different arms *l,l,l.* [57] These arms have at their inner end a small pin *i* which passing

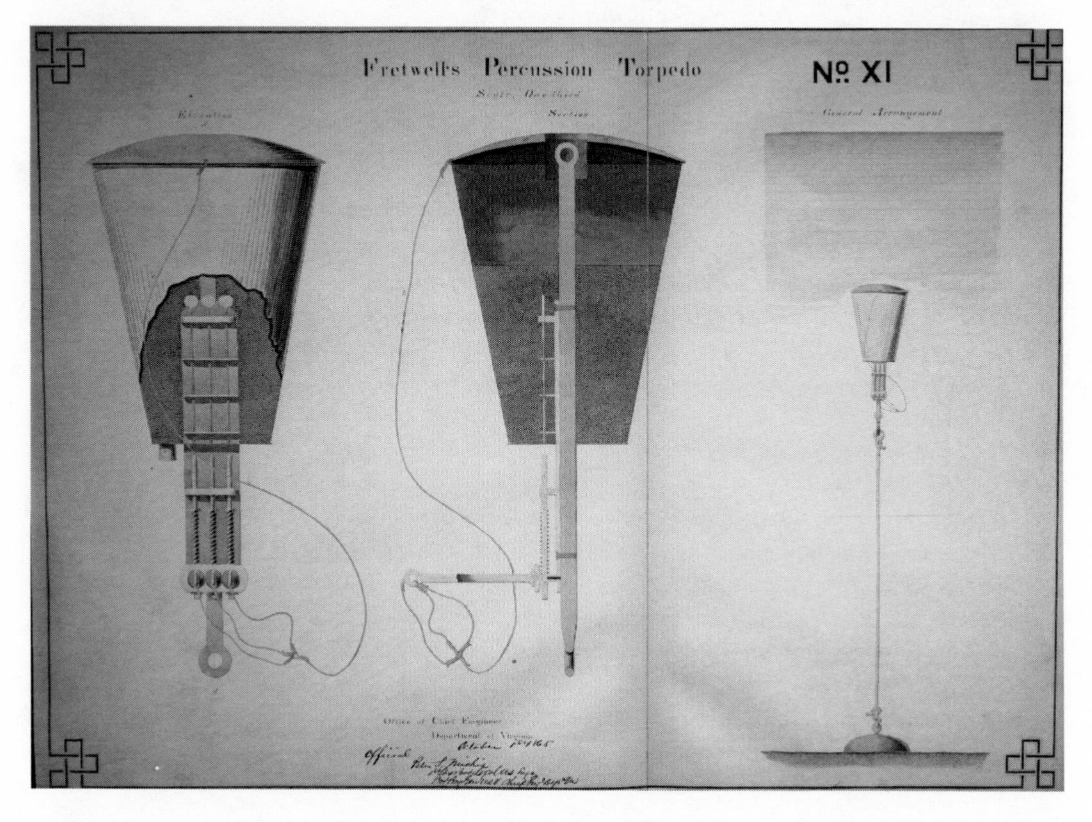

through a face plate on the frame of the apparatus enters a corresponding hole in the hammer rod (*h*) holding it down and keeping the spiral spring which surrounds it in a state of compression. Inside of the tank and in line with the hammers is a system of percussion caps *c,c,c*. Four of these caps are brought under the action of each hammer being placed on small rods and brought to a gentle but firm bearing by the thumb screws *s,s,s*. Supposing the apparatus to be set as described and properly anchored in a water course, its action when struck by a vessel will be as follows: as the cast iron plate projects slightly beyond the body of the torpedo it will first receive the blow, its inner edge [**58**] bending down or inverting over the rim which held it in position and then as the onward motion of the vessel pushes the torpedo from the perpendicular the plate tumbles off, pulling the connecting wires and through the arms *l,l,l* withdraws the pins *ii* thereby releasing the hammers which, forced by the spiral springs, fly upward and through the bottom of the tank administer a [*sic*] upward blow to the percussion caps inside causing them to burst and explodes the torpedo. The eye *o* at the upper end of the iron stem which traverses the torpedo centrally is for the purpose of lowering it gently to its position in the water and preventing the sudden jerk which would occur on striking the bottom if allowed to descend by simple gravity thus [**59**] increasing the liability of displacing some of the trigger pins or throwing off the cast-iron cap-plate. Thus lowering is done with an iron rod, whose end is best as shown in the annexed side view, is hooked into the eye. This enables the rod to be easily disengaged afterwards.

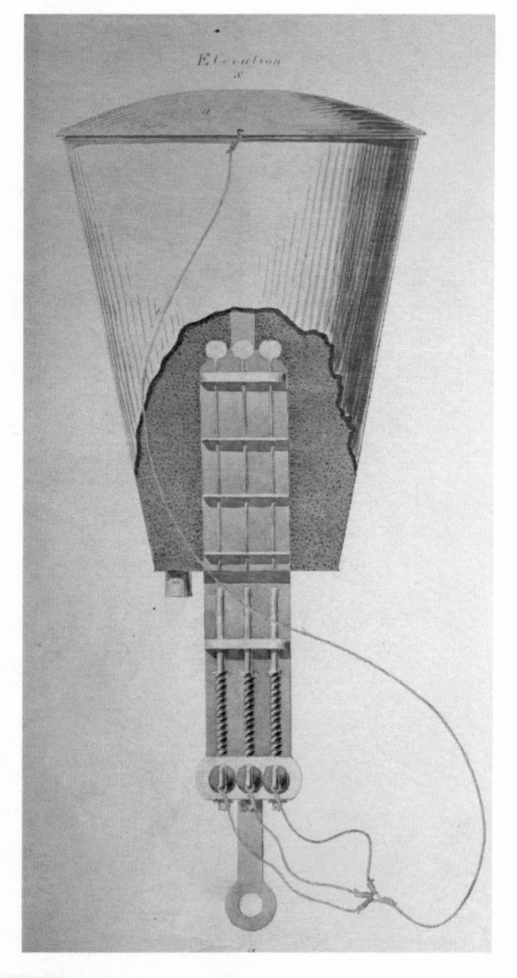

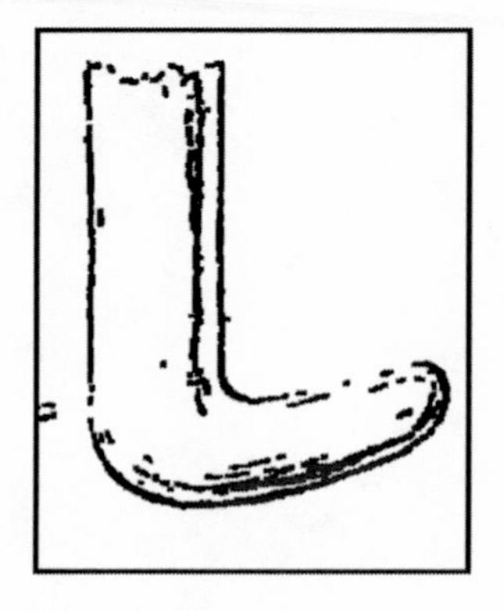

As an additional precaution against accident, safety pins are inserted in the holes *n,n,n*, which catching on the cross immediately above them, would prevent the hammers from reaching the caps should they be accidentally discharged. The pins are withdrawn after submergence by means of a line 100 feet in length which is sufficient to insure safety.

This torpedo is among the [**60**] earliest used and continued to be employed to the last. While it seems almost certain of explosion when struck by a vessel and has probably been the most successful one in use, it is still defective on account of the facility with which it may be exploded by the line drag described in the article on Swaying Boom Torpedoes, or by very strong water in time of freshets.

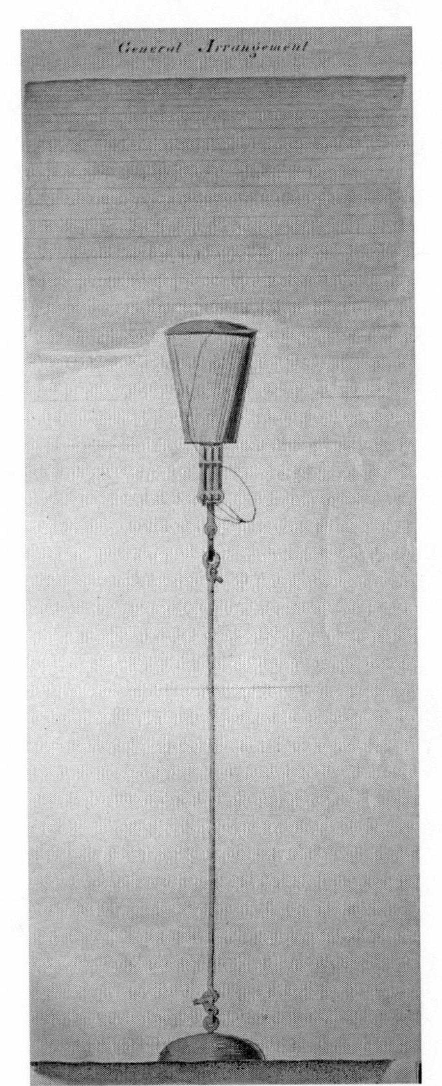

The name of the inventor is Singer, that which it bears being due to its being principally in the hands of Dr. [John R.] Fretwell of Texas who was personally engaged in planting most of them.

Plate No. XII
Shell Torpedo

This torpedo was used for the **[61]** defense of obstructions in rivers and harbors. They were bolted in an inclined position to a timber frame work which was sunk upon the obstruction and loaded with stone. Their purpose was the destruction of vessels running upon the obstructions and to render their removal more difficult.

Attached to a frame work as shown in the sketch given below they were sunk across the entrance of slips in wharves. Fixed in this manner numbers of them might be seen by close scrutiny at low tide **[62]** just below the surface of the water in [the] harbor of Charleston.

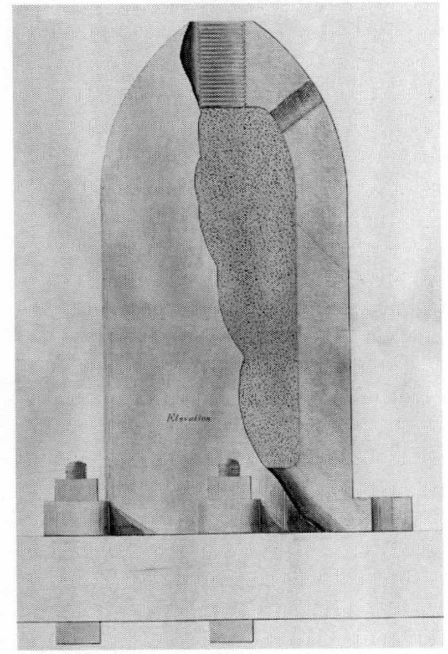

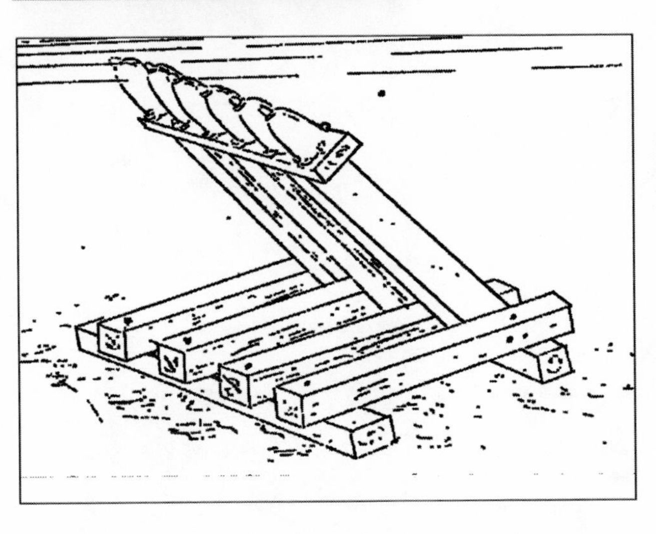

Plate No. XIII
Current Torpedo

The mode of operating this torpedo is to attach it to the end of a line about 250 to 350 feet long, the other end of which is fastened to a log of wood or "dummy."

With the line stretched transversely to the direction of the current the torpedo and dummy are floated down on the vessel it is designed to destroy so arranging that its bow

shall hang the line as near its center as possible. Knots are made along the line to prevent either the torpedo or dummy from dragging the other around and floating off.

Hanging across the bows of the [63] vessel the torpedo swings around against it sides and being held stationary by the line, the current acts on the propeller wheel causing it to rotate. The rotation of the wheel carried it from under the detent which dropping releases the hammer that falls on the cap and explodes the torpedo.

Plate XIV
Current Torpedo

This torpedo is shown in plate No. XIV. It differs from the one immediately preceding only in mechanical details and in the use of gun caps instead of percussion primers, the mode of action being the same in both cases.

By means of the worm and gear wheels the propeller wheel turns the hammer until it brings a slot on its inside into coincidence with a [64] rib on the stem when it falls, exploding caps and torpedo.

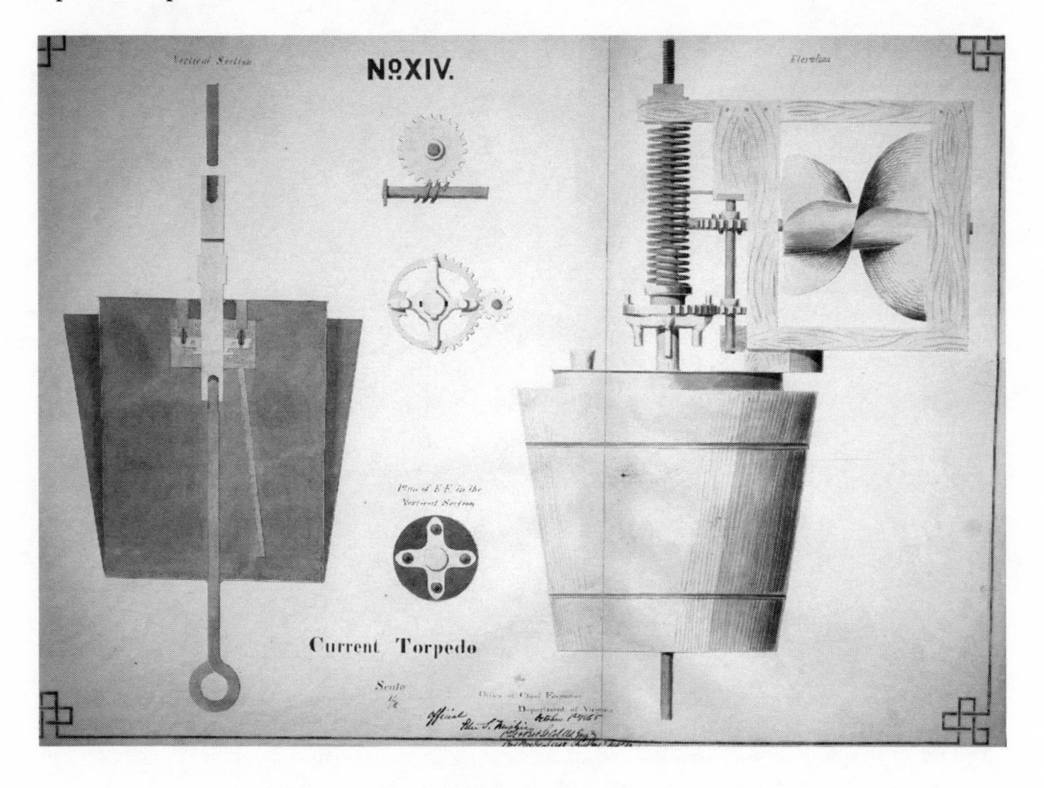

Plate XV
Hydrogen Torpedo

This torpedo which is shown in Plate No. XV depends for explosion upon the well known phenomenon of rendering spongy platinum incandescent by throwing on it a jet of hydrogen gas.

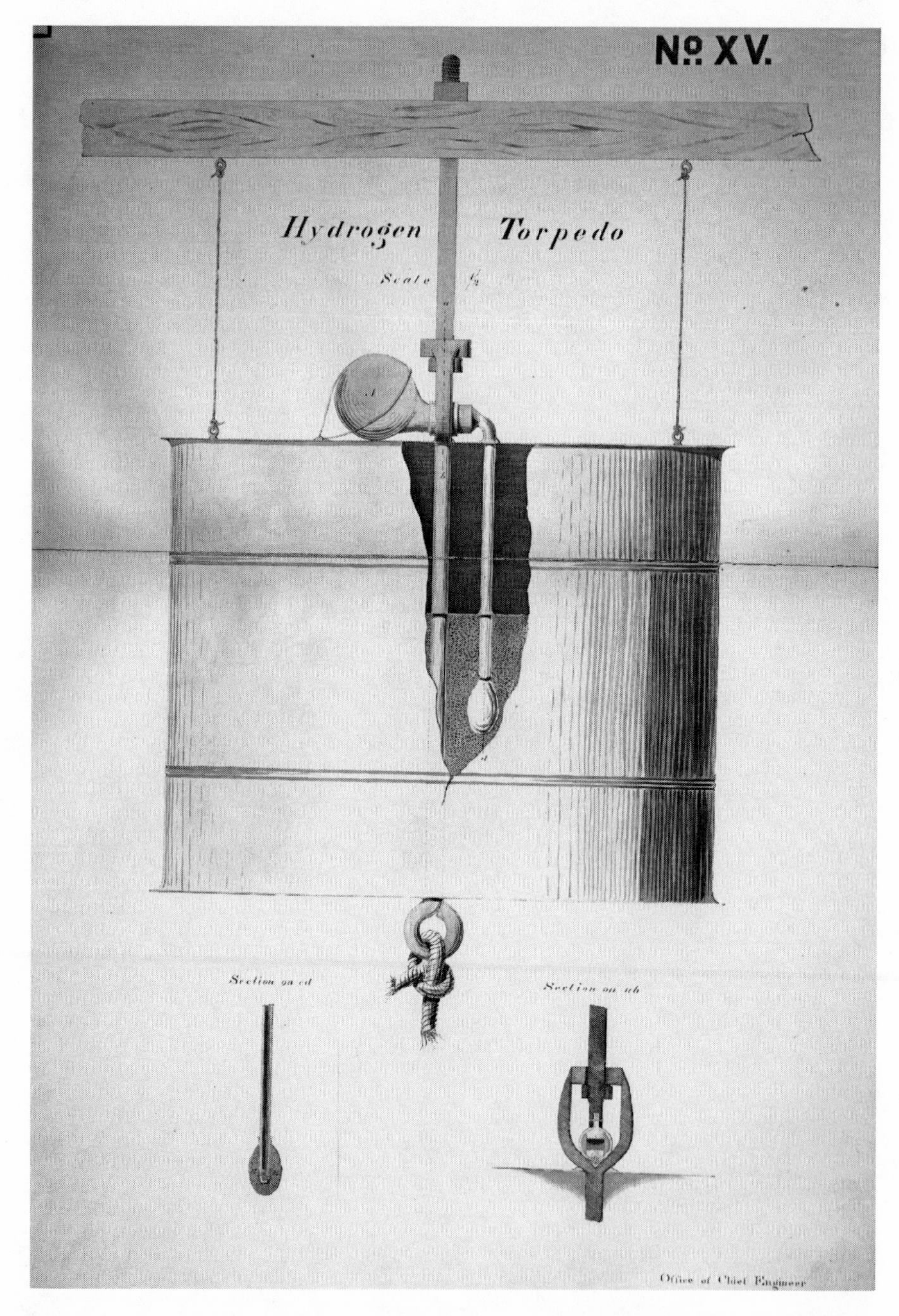

The principle is applied as follows: The beam across the top is fastened rigidly to the upper part of the stem which by means of the neck and collar, shown at its lower end, is capable of turning independently of the torpedo. The lower end of this stem is connected with a ground joint cock as is shown in the section on *ab*. The small cords are for preventing the arm from [66] turning accidentally, but should the beam be struck by a vessel they would be ruptured, the stem turned and the cock opened.

This allows the hydrogen gas of which there are several atmospheres in the globe *A*, to flow through and down the pipe *B* (see section on *cd*) where it meets with the spongy platinum *c*, which is raised to a high heat and explodes some fulminate immediately surrounding it, thence the rifle powder *E* and torpedo.

Plate No. XVI
Ground Torpedo

This torpedo consists simply of an ordinary shell with an exploding arrangement fixed in the fuse hole.

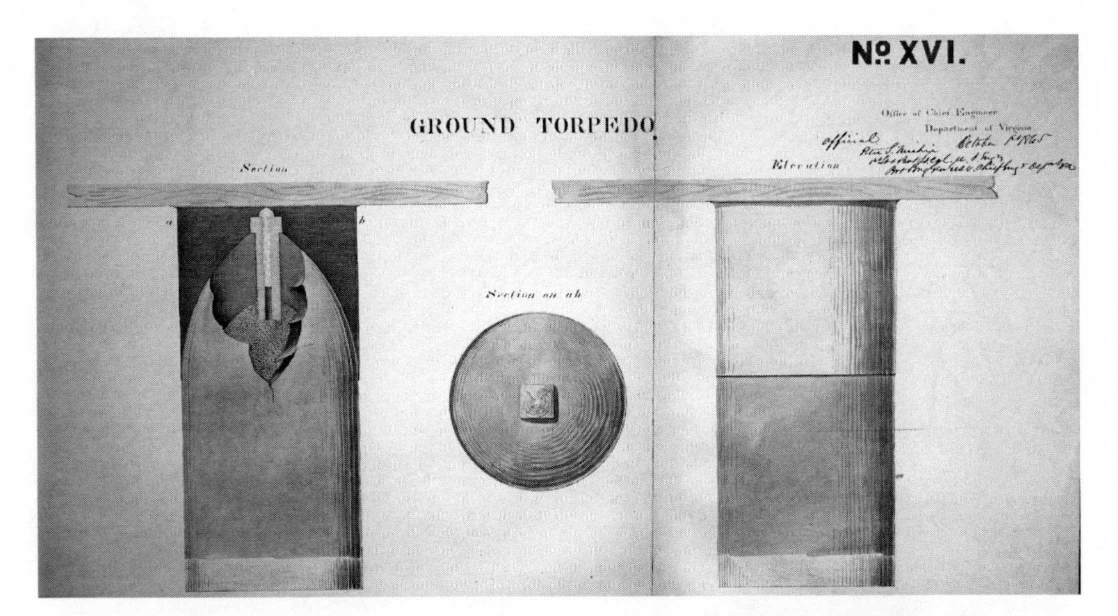

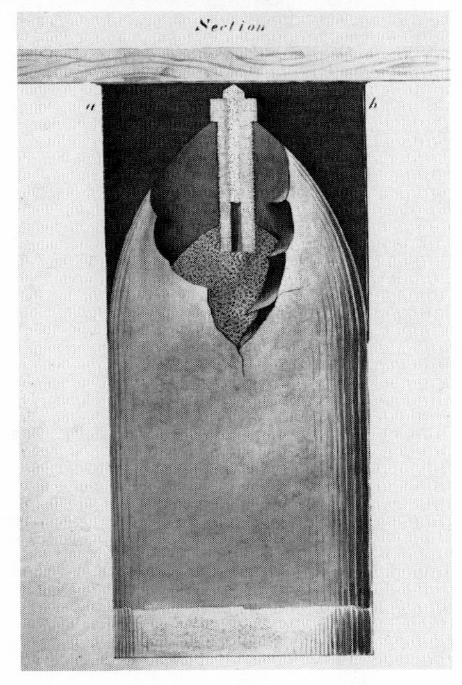

The exploder is composed of a sensitive primer inserted in a wooden plus and with an inverted tin cap, placed over [66 {*sic*}] and in contact with it. The whole was buried in the ground so that the tin cap would be an inch or two beneath the surface and carefully covered with earth so as to conceal it from observation.

Frequently several were placed in a row and piece of board laid across so that the chances of explosion would be increased.

A pressure of seven pounds would burst the primer, so that the weight of a man would certainly cause an explosion.

They were planted in roads and in front of fortifications to repel a charge. This is the torpedo which was buried in such large numbers between the lines of abatis and chevaux-de-frise in front of the rebel defenses near Chaffin's Bluff on the north side of the James River. The [67] shells used were generally condemned shells.

Plate No. XVII

Horological Torpedo

This as its name implies, is an arrangement by which the explosion of the torpedo is effected after the lapse of a certain time, a point which it is often necessary to attain in operations of this character.

The exploding device consists simply of a train of clock work which turns the wheel w fixed on the axis of the mainspring. The rotation of this wheel brings the recess m, under the pin d, which dropping into it lets the lever arm *a* descend and releases the hammer h, which being impelled by the spiral spring above it strikes and explodes the cup c, and fires the torpedo.

[**68**] This is the torpedo which was used to destroy the ordnance boats and wharf at City Point in 1864. It was arranged as shown in the plate, in a small wooden box, which contained some twelve or fifteen pounds of powder.

The clock work was put in motion by a wire which passed through the side of the box and rested on the balance wheel of the clock movement.

The individual who was charged with this project landed at City Point in a small boat. He was dressed in citizen's clothes and carried the box containing the torpedo on his back. He succeeded in passing the sentinel on duty at the ordnance wharf and deposited the torpedo on one of the barges, and he was but ¾ of a mile away before the explosion took place.

[69]

Plate No. XVIII
Horological Torpedo

This torpedo is for the same purpose and is operated in the same manner as the one preceding. The mechanical means by which the explosion is accomplished being varied in some degree.

The motion of the clock work, through the intervention of the lever which hooks on the pin of the mainspring wheel turns the T-shaped vertical lever from under the arm which rests on it, which descending allows the detent, whose toe is caught in the notch of the stem of the piston, to slide back and release the piston which falls on the primer beneath with a force due to the strength of the spring.

The drawing in the plate represents the torpedo as fixed for floating down on a vessel as described in the article on current torpedoes.

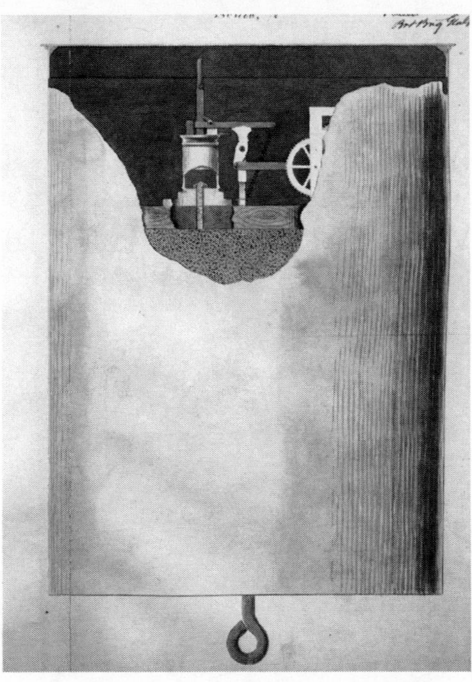

[70]

Plate No. XIX
Arm Torpedo

This torpedo belongs to the anchored buoyant class.

The exploding arrangement consists of three arms *a,a,a,* about five feet long, making equal angles with each other and let into sockets *c,c,c.*

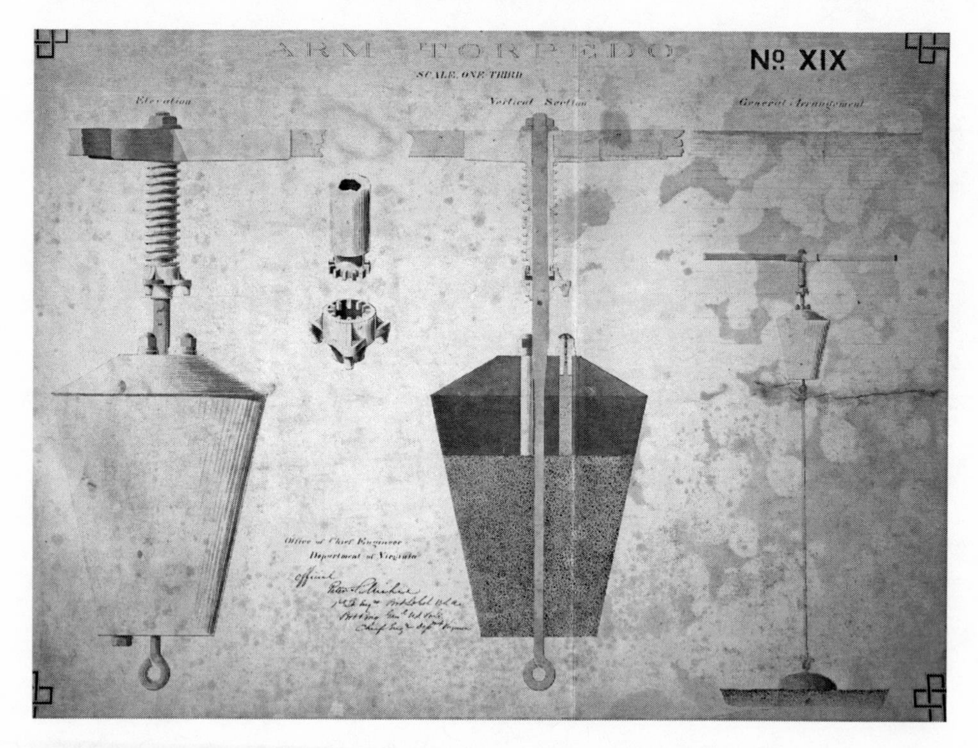

The sockets are solid with the sleeve *b*, turning on the stem *m* of the torpedo. This sleeve has at its lower end a set of cog-shaped teeth *n*, which are under corresponding teeth *i*, on the inside of the hammer *d.*

While the sleeve is free to turn, the hammer is prevented so doing by a rib on the side of the stem.

Hence if a vessel strikes the arm *a*, they are carried about, and with them the sleeve *b*, which turns the teeth *n*, from under the teeth *i*, bringing them opposite the [71] spaces and releasing the hammer *d*, which impelled by the spiral spring, falls on the primer cap beneath and explodes the torpedo.

As the teeth are only 3/16 of an inch wide, a turning of this amount in either direction is sufficient to liberate the hammer.

A safety pin *s* is used while placing the torpedo in position.

Plate No. XX
Stewart's Sight Adjusting Apparatus

This is an apparatus for the adjustment of the sights of cannon in a more accurate manner than was possible with the method of fixing by measurement on the exterior of the piece.

The exterior and interior surfaces of a gun are not concentric owing to the imperfection of the boring machinery.

During the war a number of old guns were broken up at Tredegar Iron Works at Richmond for remelting, and it was noticed among them a very great departure [72] of the center from its true position.

In one gun made at Alger's foundry, Boston, the eccentricity amounted to an inch and a half.

In a gun like this, sighted by its exterior, the line of fire could not coincide with the line of sight, except by the accident of the deflection occurring in a vertical plane.

The apparatus under remark is designed to overcome this difficulty, and was used by the rebels in adjusting the sights of all their guns during the last three years of the war.

The principle involved in it consists in a virtual prolongation of the axis of the bore beyond the muzzle of the piece, and the tracing backward over the exterior of the gun a line parallel to this prolongation.

It consists of a hollow, cylindrical shaft *s*, which extends down the bore of the gun to be sighted, and whose axis is made exactly coincident with that of the bore by [73] means of expanding heads *A* and *B*. Each of these heads is composed of four segments, as shown in the sectional drawings, resting on cones *I*, and *C*, which expand these segments equally in their respective directions when moved along their larger bases by the milled-head screws *G* and *D*, which latter operates through the intervention of the rod *N*.

By means of set screws *m*, *m*, and heads *H*, two arms are attached to that portion of the shaft which projects beyond the muzzle. The arms have those edges which are turned

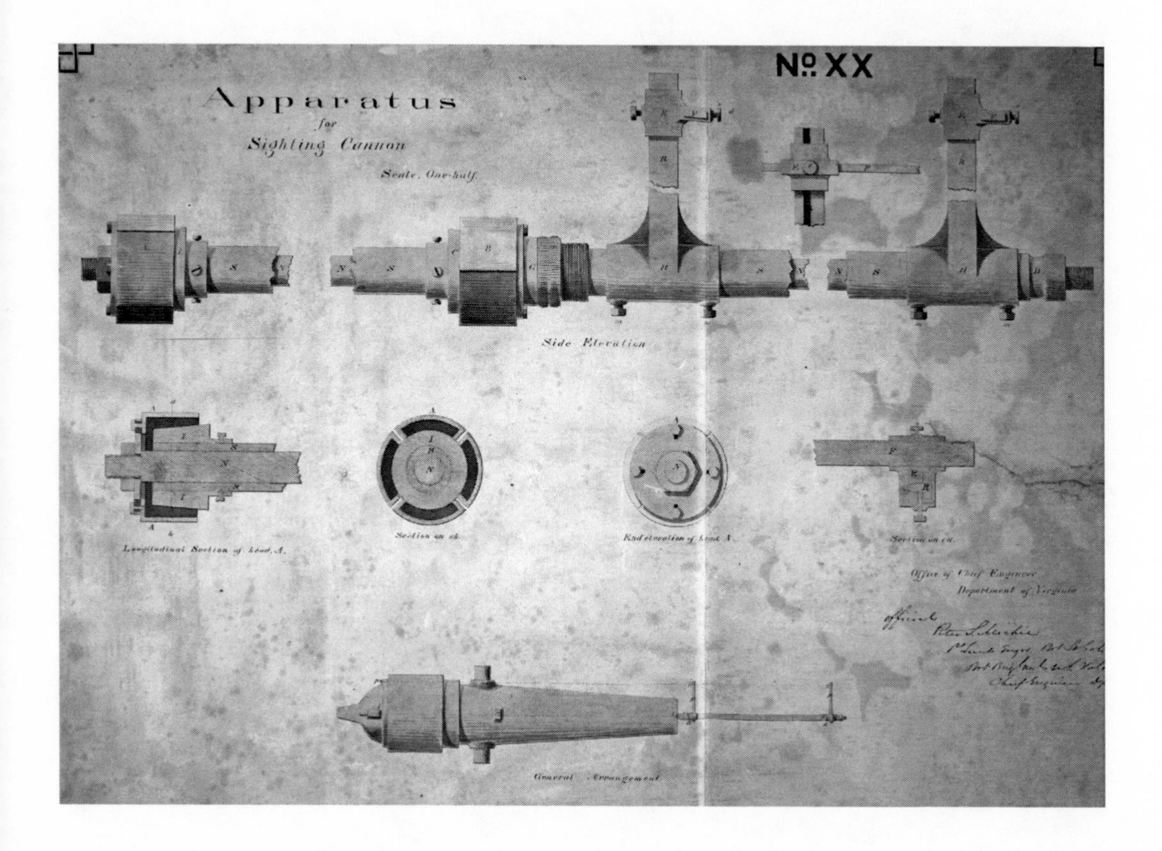

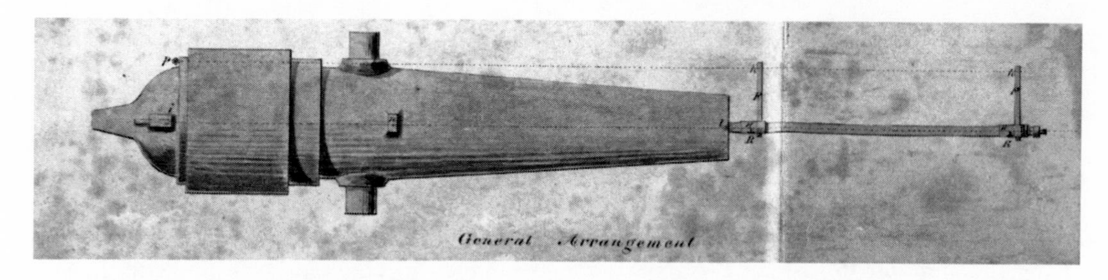

from the observer, parallel to each other and intersecting the axis of the shaft and hence in a plane passing through the axis of the bore.

To use the apparatus, the axis of the trunions of the gun to be sighted, is first made horizontal, then the heads run down the bore and expanded, and the shaft *s* turned until the edges of the arms *R,R* come into a vertical plane.

[74] A fine silk cord is stretched touching the edges of *R,R* and parallel to the axis of the shaft, and carried back over the outside of the gun. The distance of the cord from the axis is equal to that of the summits of the sights from the center of the bore.

To facilitate the placing of the cord in this position the arms are graduated.

The line of the sights are [*sic*] brought into coincidence with that of the cord and the adjustment is complete.

In adjusting trunion sights graduated horizontal arms *F,F* sliding through heads *E,E* are employed and the silk cord stretched in position by them.

Plate No. XXI
Gardner's method of attaching the paper cartridge to the Ball

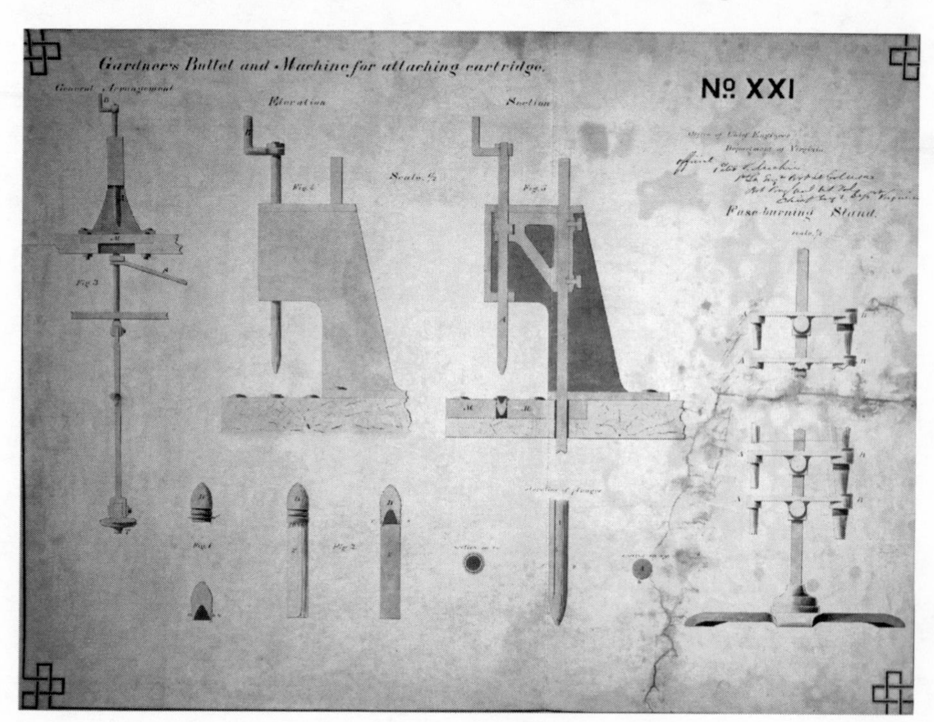

The object of this device was to expedite the manufacture of musket cartridges by superseding the necessity of tying the paper [75] to the ball and at the same time swage the ball to exact caliber and render the connection between it and the paper more accurate. It is the invention of F.J. Gardner of N.C. and was used in the rebel arsenal at Richmond.

The bullets used were cast in the form shown in fig. 1 having a circular flange a running around them immediately below the cannelures.

The paper was attached by turning down this flange upon the base of the ball, *c*, and paper *g*, being caught between them as shown at *e*, in fig. 2. This was accomplished in a very simple manner in the machine shown in fig. 3, 4, and 5.

The edge of the cartridge paper cut to proper size was inserted in the slot, *b*, of the steel plunger, *A*, and wound smoothly upon it by turning the handle *B*.

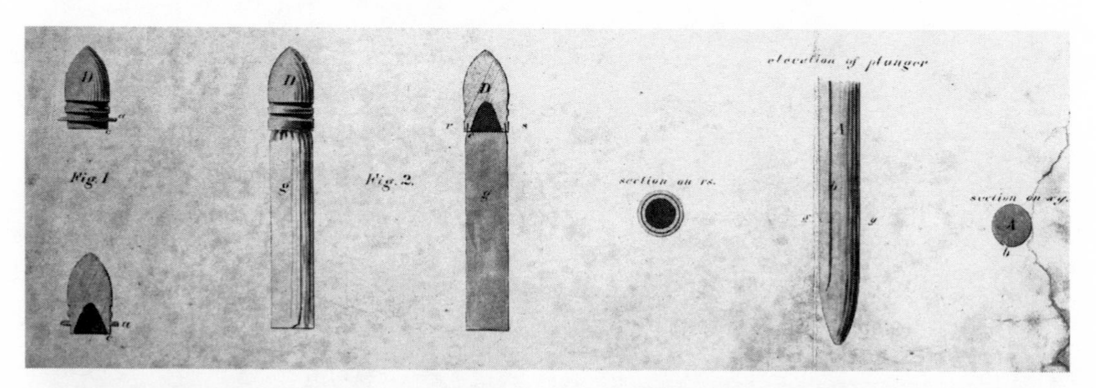

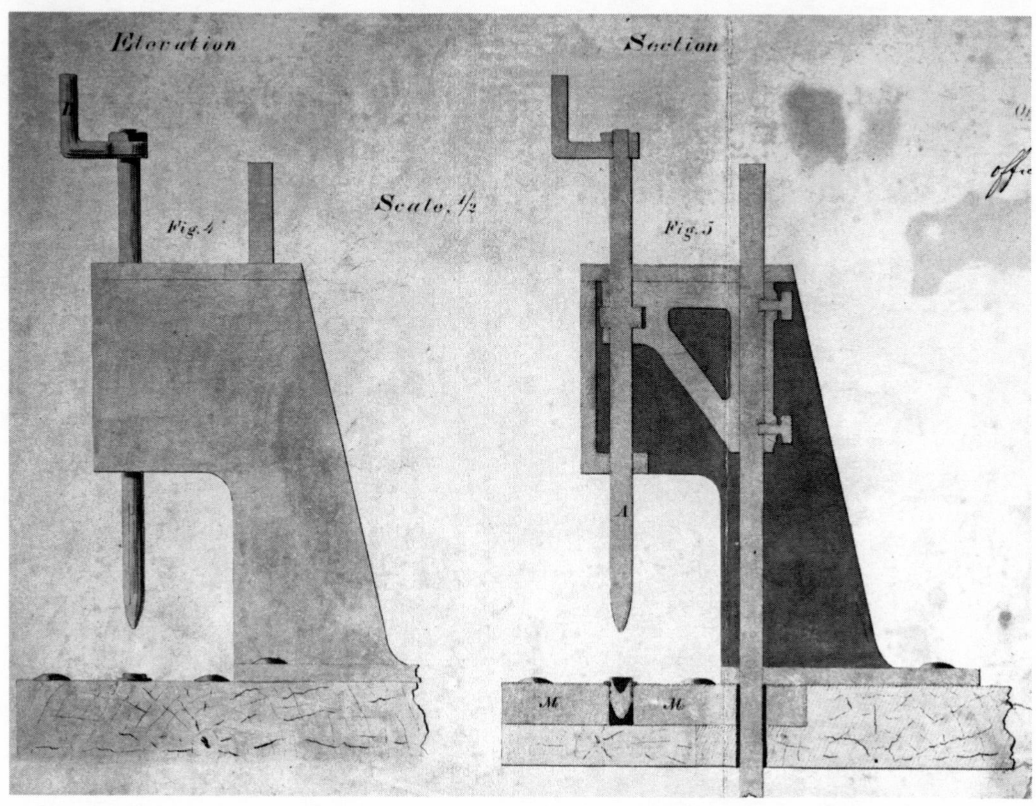

The foot is then placed on the treadle, *T*, and the plunger and paper brought down on the bullet, *D*, forcing the latter through the swaging plate *M*, the flange turned down and the [76] paper caught.

Upon removing the foot from the treadle, the spring, *S*, lifts the plunger and the machine is ready to repeat the operation.

Fuse Burning Stand

This was for the purpose of ascertaining the time occupied by a fuse in burning, with greater accuracy than is attainable in burning a single one.

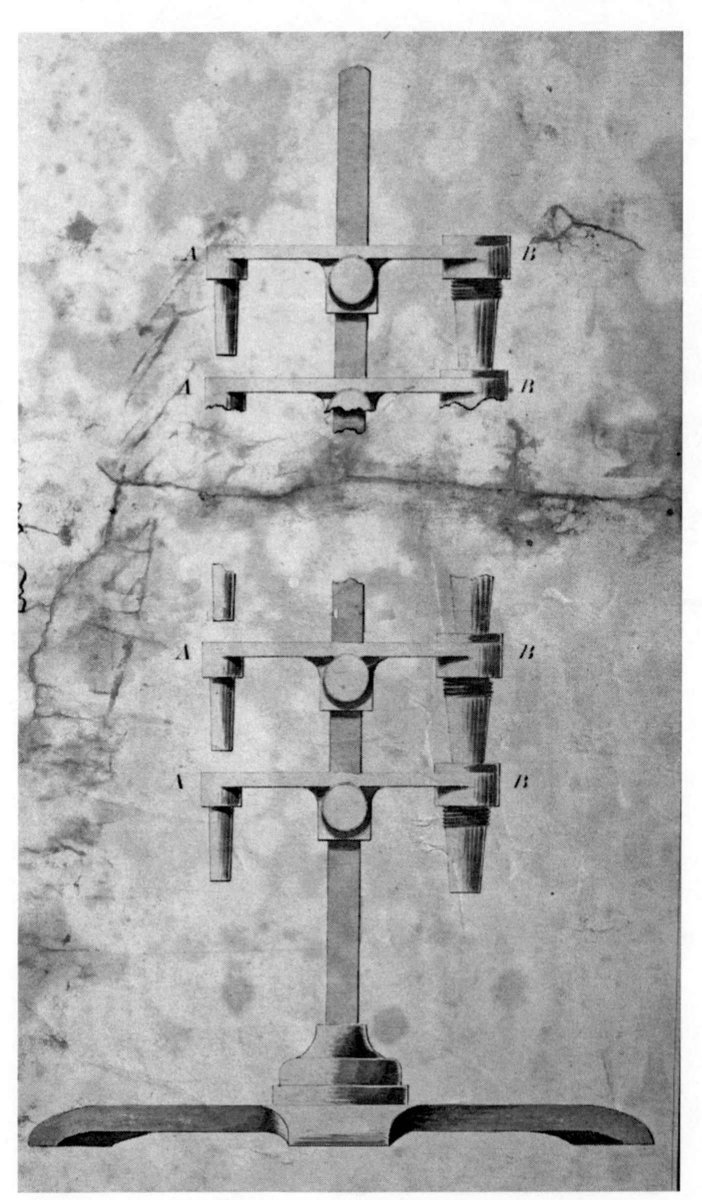

It consists of a standard *M*, supporting sockets *A,A* in which the fuses to be tested are placed.

A column of fuses in contact with each other is thus formed and the whole burned from the upper end. The time occupied by the column in burning, divided by the number of fuses composing it, will give the exact time occupied by any one of them.

The error of observation of time which would otherwise have been accumulated in a single fuse is by this means distributed among the [77] whole of them. Ten was generally the number burned which gave the time decimally without the necessity of a division. The fuses were separated by a slight interval to enable the observer to see when the burning point of the column passed from one to the next below and know when to turn these burnt out, aside, in order to facilitate the egress of the flame, and render the burning more uniform.

When Alger's fuses were to be tested under water caps, the plugs, with the fuses fixed as for service, are screwed into the sockets *B*, and burned as described.

Operating with floating torpedoes

It is best to use two boats. Let the dress of the crews and everything about the boats be of a lead color. The boats must mount about four muffled oars each. One carried the torpedo, the other the dummy with the connecting line attached. Approach within a safe distance of the vessel to be destroyed, say 250 yards.

Get in such a position that her masts will [78] be in line. As she is free to swing this will indicate the direction of the current.

The boats now separate and move slowly in opposite directions, until the line connecting the torpedo and dummy is tightened. They are thus placed noiselessly in the water.

Whatever safety arrangement is used in the torpedo is released, and if a horological one, the clock work put in motion, and the torpedo started on its errand. The boats then get away with all possible expedition.

Communication between the boats must be had only by signs, and their crews should be very thoroughly exercised in their duties before any actual operations attempted.

Peter S. Michie

Richmond
December 1865

P.S.Michie
Bvt. Brig.Gen. U.S. Vols.
Chief Engineer
Dept. Va. & N.C.

Editor's Appendix 1: Vessels Sunk or Damaged by Confederate Torpedoes[1]

February 14, 1862: *Susquehanna*'s launch receives minor damage on the Wright River, Ga.

In February 1862, as preparations for the isolation of Fort Pulaski continued on the Savannah River, prior to the siege and capture in April 1862, Federal vessels discovered five "infernal machines." The five containers were floating in the Savannah River across the entrance to Wright River. They were constructed of metal cases which served both as "air chambers" or "buoys," and each contained approximately 70 pounds of gunpowder. They were anchored so as to be covered at any stage by the tide when the bar was passable for vessels, but they were visible at low tide.

Naval Lieutenant Commanding John P. Bankhead snagged one and brought it aboard the gunboat *Unadilla.* Because the torpedo mechanism was unknown, naval squadron commander John Rodgers had the device placed on the shore and rifle fired into it, which caused it to explode. The internal "machinery" was recovered intact, revealing that inside the powder chamber was a friction primer like those used to discharge cannon. The primer "was arranged so as to ignite the 'blowing-up' charge upon the pulling of a string. This string was tied to a wire coiled up on the head of the buoy; the coil of wire was to be drawn out by the impact of a passing gunboat."

Later that night, one of the *Susquehanna*'s launches was towing a barge carrying artillery to Venus Point on Jones Island in the Savannah River, the site where Battery Vulcan was being erected. The launch had traveled about 200 yards above the line of torpedoes, when it was damaged by an explosion. Commander Rodgers thought this explosion might have been caused by a galvanic torpedo connected to Fort Pulaski, and ordered Lieutenant Bankhead to "sink the remaining machines with rifle shots" which he proceeded to do.[2]

Diagrams of the torpedoes pulled from the Savannah River[3] (see following page).

December 12, 1862: Armored river gunboat *Cairo* sunk on the Yazoo River, Mississippi

The U.S.S. *Cairo* was a steam powered 512-ton gunboat built under contract by James B. Eads & Co., and was placed in service in 1862. She was armed with eighteen guns.[4]

At 7:30 A.M. on December 12, 1862, Captain Henry Walke directed a group of light-draft gunboats (U.S.S. *Pittsburg, Marmora, Signal* and *Cairo*) and the ironclad ram *Queen of the West* to ascend the Yazoo River from their anchorage above Vicksburg. Walke had a double mission: destroy Confederate shore batteries and destroy the floating torpedoes, several of which had been spotted the previous day. The *Queen of the West* was to protect the *Marmora* and *Signal*, which were assigned to job of finding and destroying torpedoes.[5]

At the Confederate shipyard up river at Yazoo City, Confederate Captain Isaac N. Brown was supervising the construction of three ironclads. Zere (or Zedekiah) McDaniel and Francis M. Ewing, members of the Confederate Submarine Battery Service, were busy manufacturing torpedoes. Their device of choice was the 5 gallon wicker-covered demijohn torpedo. The firing mechanism was the cannon friction primer connected by wire, through a water-proof seal of gutta percha and plaster, to the exterior, where it as attached to a wire stretching from bank to bank. When the wire from the demijohn was pulled, it ignited the primer, which in turn, detonated the gunpowder within the demijohn container.[6]

With the *Marmora* in the lead, the Federal vessels cautiously ascended the Yazoo River. When the fleet was approximately sixteen–eighteen miles from the mouth of the Yazoo River and four or five miles below Haynes' Bluff, Mississippi, they came under heavy musket fire. As the *Marmora* attempted to negotiate a tight bend, the *Cairo* ascended to join her. The *Marmora* opened on what seemed to be a floating block of wood. Lieutenant Commander Thomas O. Selfridge, Jr., commander of the *Cairo*, sent off a boat and retrieved what he thought was a part of an exploded torpedo.[7]

In the meantime, Ensign Walter E.H. Fentress, in a boat launched from the *Marmora*, saw a line in the water which he cut and torpedo bobbed to the surface. Fentress then found another line, which he pulled and towed a torpedo up and onto the shore. There, he began to open the torpedo.[8]

While the small boats were searching for torpedoes and pulling lines, the *Cairo* had

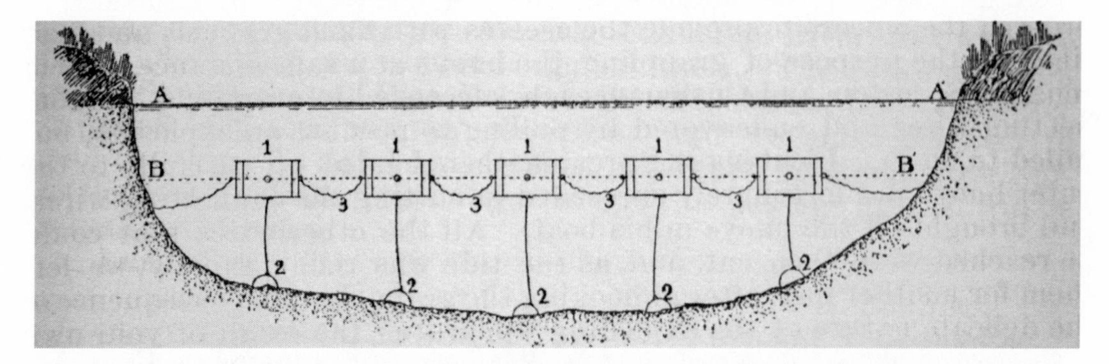

Torpoedoes pulled from the Savannah River: T*OP: Section of torpedo:* ***a b c d****, water-tight tin case;* ***e e****, air chamber; P powder chamber containing 70 pounds pf powder;* ***f****, ordinary cannon friction primer, held at center of charge by strip of tin,* ***i m****; attached to it is the wire* ***g f****, passing through a box,* ***h****, filled with wax and tallow;* ***m o****, mooring line.* B*OTTOM:* ***1, 1, 1, 1, 1****, torpedoes;* ***2, 2, 2, 2, 2****, anchors;* ***3, 3, 3, 3, 3****, spiral wires connecting torpedoes;* ***A', A'****, high water mark;* ***B', B'****, low water mark.*

drifted towards the river bank. Now, between 11:30 A.M. and noon, she and the *Marmora* slowly resumed their ascent of the Yazoo. Suddenly, two violent explosions blasted beneath the *Cairo*, one on the port side and the other under the port bow. As the *Cairo* had pressed against the torpedoes, the vessel had pulled the torpedoes onto their side, pulling and igniting the primers, and exploding the gunpowder against the hull. The force of the explosion raised cannons from the *Cairo*'s deck. The vessel's frame was shattered and within five minutes she was below water. Fortunately no one was killed.[9]

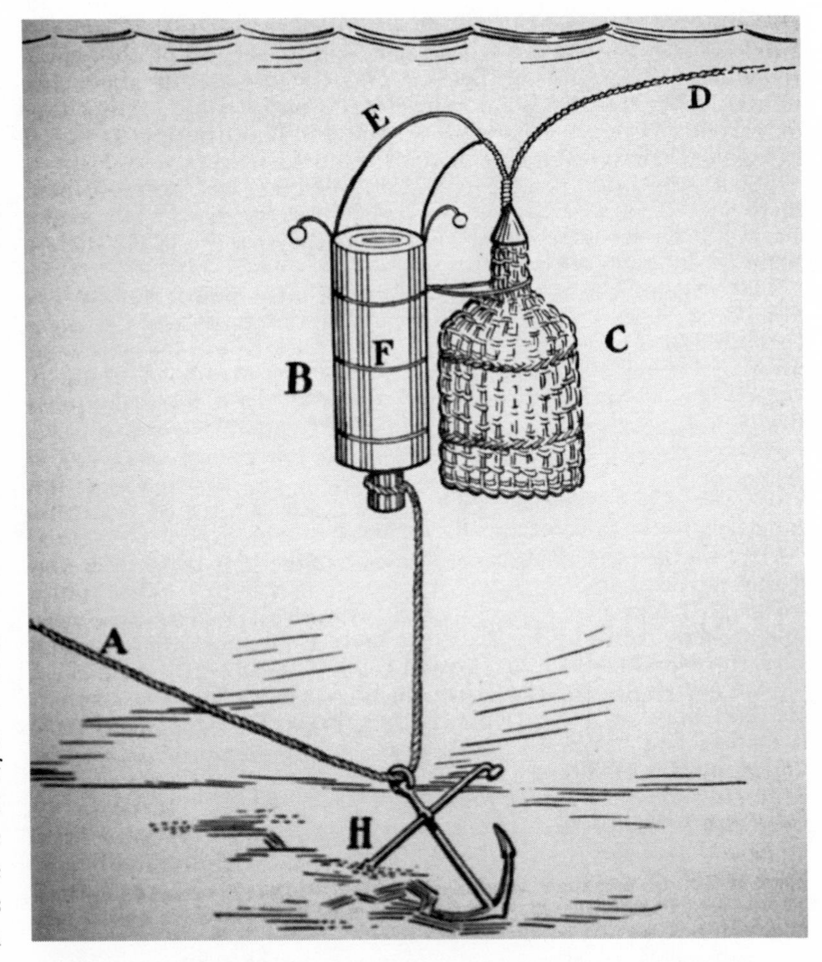

A, the rope by which the torpedo was sunk. ***B****, a log of wood attached to the demijohn* ***C*** *(containing the bursting charge) by the bands* ***F. D*** *and* ***E****, wires running into the vessel containing bruising charge.* ***H****, weight by which the engine was kept under.**

The *Queen of the West* rescued the *Cairo*'s crew. The ram then pulled away the *Cairo*'s stacks, all which had remained visible of the vessel, to prevent the Confederates from later finding her. Federal small boats in the river retrieved an additional ten or twelve torpedo as the Union vessels shelled the shore. Federals located the Confederate torpedo manufacturing facility on shore and destroyed it.[10]

An unsuccessful salvage attempt on the *Cairo* was made in 1864. The *Cairo* was located again in 1959 by Edwin C. Bearss, then historian at Vicksburg National Military Park, and the pilot house was retrieved in 1960. The vessel itself was recovered in 1964 and she is preserved and now on display at the park.[11]

Federal naval officers submitted the diagram of the torpedoes they encountered (see above). Although the configuration is somewhat incorrect, it does show the demijohn torpedoes anchored to the shore by wires. (These anchoring wires were incorrectly interpreted by the Federal authorities as part of a galvanic firing mechanism.[12]

**ORN*, Series 1, vol. 23, pp. 548–549.

January 14, 1863: River gunboat *Kinsman* receives minor damage on Bayou Teche, Louisiana

This 245-ton side-wheel steamer was built in 1854 as the steamship *Grey Cloud.* She was seized in New Orleans in May 1862 by Major General Benjamin F. Butler's forces, fitted out by the army as a river gunboat, and renamed the *Colonel Kinsman.* She was transferred from the army into the navy on January 1, 1863.[13] Although her armament is not recorded, the other gunboats accompanying her in January, 1863, were each armed with five guns.[14]

Brigadier General Godfrey Weitzel learned from various sources that the Confederates, with the steamer *J.A. Cotton*, were planning an attack on his forces at Berwick Bay, Louisiana. He decided to initiate a preemptive attack to destroy the gunboat. Lieutenant Commander Thomas McK. Buchanan led the naval component: gun boats *Calhoun*, *Estrella*, *Kinsman,* and *Diana.* Acting Volunteer Lieutenant George Wiggin commanded the *Kinsman.*[15]

The expedition set out on January 13, 1863, with the vessels transporting the artillery, cavalry, and six infantry regiments to Pattersonville. By 3:00 P.M. the force had reached Bayou Teche and encamped. On the morning of January 14, the combined force began moving on the Confederate positions, with the *Kinsman* in the lead, followed by the *Estrella, Calhoun* and the *Diana.* The bayou had been obstructed with logs and other debris. By 9:00 A.M. the first three vessels, along with the ground artillery, were firing on the *Cotton.* The *Cotton* returned fire, as did Confederates in rifle pits on the banks. Shortly after the beginning of the attack, Lieutenant Wiggin maneuvered the *Kinsman* near the obstructions and as he was "endeavoring to get out of range of the rifle pits," a torpedo exploded under the stern "unshipping her rudder" but "did no serious injury." The *Kinsman* then retired from the range of the Confederate rifle pits, and the *Calhoun* moved into the forward position. The 8th Vermont Infantry now arrived and drove the Confederates from the rifle pits, taking 41 prisoners. The Federal vessels then moved up to the obstructions, where they found another torpedo with wire leading to the shore; sailors carefully removed it. During the night the Federal vessels lay at anchor near the obstructions; dawn revealed the *Cotton,* swung across the bayou and burning, further obstructing the channel. The Federal forces then withdrew.[16]

February 28, 1863: Monitor *Montauk* seriously damaged in Ogeechee River, Georgia.

This 750-ton single-turret monitor was delivered to the Navy in December 1862. She was armed with a 15-inch and an 11-inch Dahlgren gun.[17]

In December 1862 three Federal vessels lay off the mouth of the Ogeechee River, south of Savannah, Georgia. Trapped within the river was the Confederate blockade runner *Nashville.* Although she would be bottled in the river for three months, she was well protected under the guns of earthen Fort McAllister, and a line of obstructions and torpedoes farther downstream, near the mouth of the river.[18]

In January 1863 Commander John L. Worden, previously in command of the famous *Monitor,* was dispatched in his new ironclad *Montauk* to the mouth of the Ogeechee River. He was accompanied by a squadron of five other vessels, the gunboats *Seneca*, *Wissahickon,* and *Dawn,* the *Williams,* and the tug *Adger.* Reconnaissance revealed obstructions in the mouth of the river, some of them suspicious for torpedoes. On January 27, the vessels carefully moved into the Ogeechee River, avoiding the obstructions, and opened fire on the

Nashville and the fort with little result. Withdrawing at noon and replenishing supplies, the squadron ventured in again on February 1 with similar results.[19]

Again supplied, the Federal vessels watched and waited. Around 4:30 P.M. on February 27, they saw the *Nashville* moving and feared an attempt at escape. By the next morning the *Nashville* had managed to ground herself and the Federal ironclad and three gunboats moved in for the kill. Under fire from Fort McAllister, the Union vessels incredibly avoided direct hits while pounding the *Nashville*, which burst into flames and then exploded around 8:00 A.M. The wooden gunboats then withdrew followed by the *Montauk*. At 9:35 A.M., in the Ogeechee River a few yards above Harvey's Cut, the *Montauk* hit a frame torpedo and began taking water. The pumps could not keep her afloat and the pilot ran her onto the muddy bank. Inspection subsequently revealed a ten foot long crack in the hull and a three by five foot area of disarrayed iron plates, the direct area of explosive impact. A patch made of boiler iron, and full use of the pumps, allow the ironclad to be towed to Port Royal, South Carolina, for repairs.[20]

March 14, 1863: Screw sloop *Richmond* receives minor damage at Port Hudson, Louisiana.

The U.S.S. *Richmond* was a 2,700-ton wooden screw steamer launched in January, 1860. Initially armed with sixteen guns, her gun total and configuration would vary slightly during the war.[21]

On the morning of March 14, 1863, Rear-Admiral David G. Farragut assembled his fleet to ascend the Mississippi River to pass above Port Hudson and join Major General Ulysses S. Grant at Vicksburg. The *Richmond* was the slowest of his vessels, and Admiral Farragut ordered it lashed to the side-wheel steamer *Genesee*, his fastest vessel. The plan was to ascend above Port Hudson in the morning darkness. The flagship *Hartford*, lashed to the *Albatross*, led the way, followed by the *Richmond* and *Genesee*, then the *Monoghalela* lashed to the *Kineo*, and lastly followed by the side-wheeler *Mississippi*.[22]

Captain James Alden, commander of the U.S.S. *Richmond*, struggled to pass the batteries of Port Hudson on the night of March 14-15, 1863, along with the other vessels of the squadron. She and the *Genesee* were second in line, when a shot struck the *Richmond*'s steam pipe, in the vicinity of the safety valves, which "upset" them and let off steam. Although the *Richmond* maintained fire against the Confederate batteries for two hours, the lack of steam power, and hence propulsion, eventually told. Although the *Richmond* was lashed to the *Genesee*, the latter was unable to provide the necessary force to make headway and overcome the current. The paired ships turned back and anchored out of the range of Confederate guns.

Captain Alden concluded his report of the action with "Just before the accident to our steam pipe, a torpedo was exploded close under our stern, throwing the water up 30 feet, bursting in the cabin windows and doing other unimportant injury."[23]

April 6, 1863: Confederate steamer *Marion* sunk in the Ashley River, South Carolina.

The *Marion* was a 255-ton side-wheel steamer, which had been built in Charleston, South Carolina, in 1850. She had been taken over by South Carolina forces during the Fort Sumter crisis as a guard boat. *The Marion* was the vessel which took Federal soldier's families from Fort Sumter to New York prior to the bombardment in 1861.[24]

On April 6, 1863, she drifted from her anchorage at the mouth of the Ashley River and before anyone noticed, had scraped her bottom against a frame torpedo which she had recently positioned, hitting two of the iron projectiles. She blew out the bottom of hull and sank in less than a minute in thirty feet of water near Wappoo Creek. Captain John Flynn was killed and her machinery destroyed.[25]

April 7, 1863: Confederate steamer *Ettiwan* sunk in Charleston Harbor, South Carolina.[26, 27]

The *Ettiwan*[28] was a side-wheel steamer built in Charleston, South Carolina, in 1834. During the Civil War she transported cargo between the harbor forts and the city. A loose "keg" torpedo drifted into her and exploded on April 7, 1863 and she was run ashore near Fort Johnson. She was repaired and later wrecked on June 7, 1864.[29]

On August 30, 1863, the *Ettiwan* was tied to the dock at Fort Johnson. Without notice, she moved off and her ropes became entangled with the submarine *Hunley*, which was also tied to the dock with her hatches open. The *Ettiwan* caused the *Hunley* to capsize and the five men aboard the submarine could not escape through the small hatches and drowned.[30]

Raised and refitted by the United States Army after the war, The *Ettiwan* was used to help clear Charleston harbor.[31]

April 7, 1863: Monitor *Weehawken* receives minor damage in Charleston Harbor, South Carolina.

The U.S.S. *Weehawken* was an 844-ton single turret monitor, which was launched in November 1862. She was armed with 15-inch and 11-inch Dahlgren gun.[32] She was stationed off Charleston in the spring of 1863.[33]

The Confederates had been experimenting with both percussion and galvanic marine torpedoes for the defense of Charleston Harbor. Since April 4, 1863, the Confederates had observed "that the turrets of the far-famed Monitor gunboats were looming up against the southeastern horizon." On April 7 nine Union monitors, led by the *Weehawken*, and including the *Montauk, Keokuk, Nahant, Nantucket, Catskill, Passaic, Patapsco*, and *New Ironsides* entered the harbor. The *Weekawken* was pushing a 50 foot "devil," a device to rake up and detonate torpedoes. She moved toward a line of obstructing torpedoes which crossed the channel. She detonated one which lifted her slightly but caused no appreciable damage.[34]

During the ensuing shelling, the *New Ironsides* moved out of cannon range but unknowingly directly over a previously placed large galvanic torpedo. This device was fabricated from an eighteen foot long boiler three feet in diameter. It contained 3,000 lbs. of gunpowder and could be ignited by an electrically discharged fuse containing fulminate of mercury, located in the center of the boiler's gunpowder. Confederate signal operators in Battery Gregg and Battery Wagner waved flags, indicating that the *Ironsides* was over the torpedo. Confederate Captain Langdon Cheves closed the contact switch but nothing happened. Everything seemed in order, and the switch was closed again and again, but nothing happened. After the *New Ironsides* had slowly moved away, investigation of the retrieved mine reveals it to be intact. Inspection of the wires from the battery house revealed a deep wagon wheel track over one of the severed wires.[35] [See Rains' explanation, pp. 51–51.]

During this naval incursion, the *Keokuk* was sunk by Confederate artillery fire; the other monitors were heavily damaged, but left under their own power.[36]

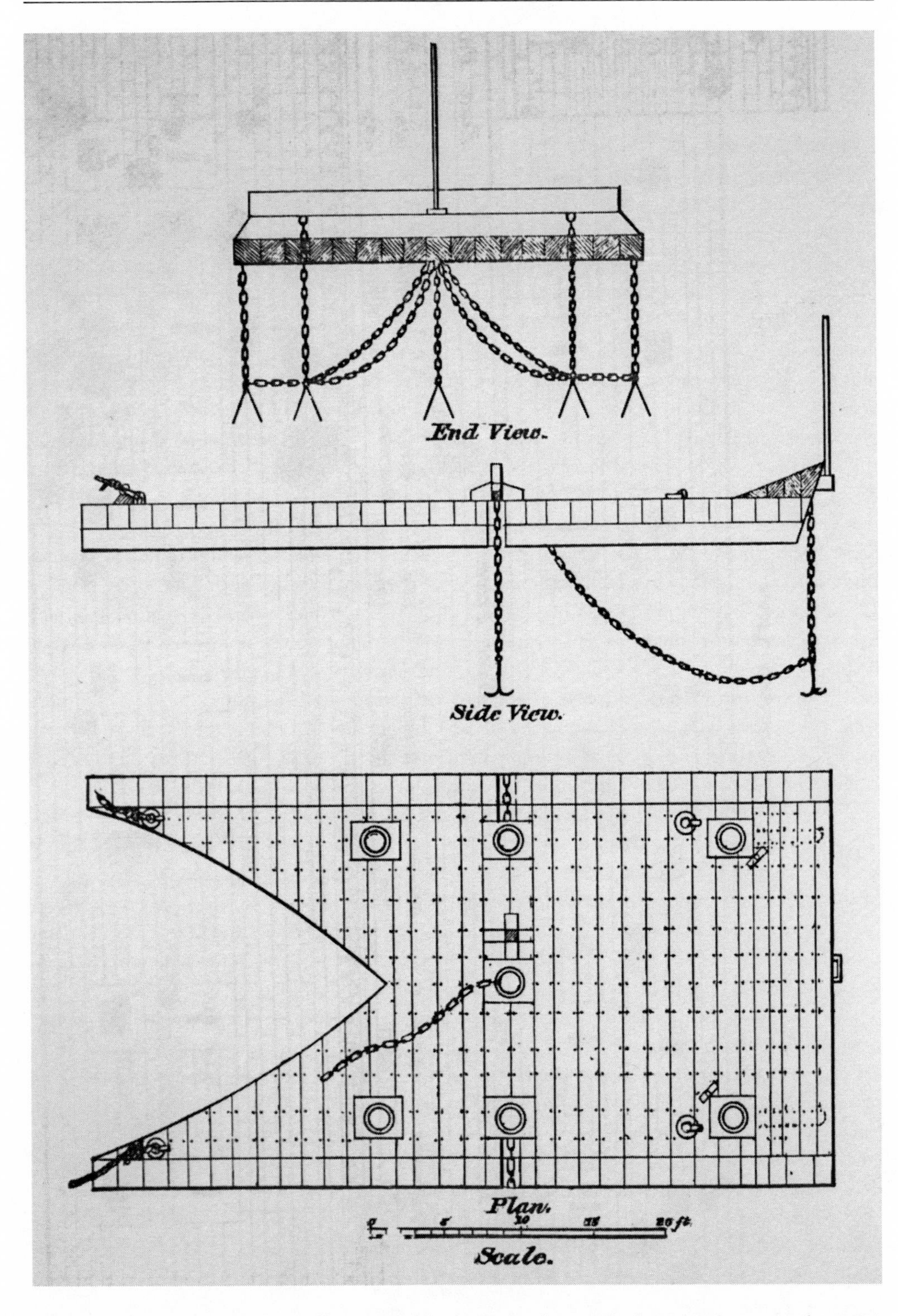

*Sketch of the Devil, or Torpedo Searcher, Carried by the Weehawken in Charleston Harbor on April 7, 1863**

**OR*, Series 1, vol. XIV, p. 254.

July 13, 1863: Armored river gunboat *Baron De Kalb* sunk on Yazoo River, Mississippi.

This vessel began life as the *St. Louis*, built by James Eads in 1862. She was refitted as an ironclad in mid 1862. She was a 512-ton vessel and was armed with 14 guns.[37]

On July 11, 1863, Major General Grant ordered Admiral David D. Porter to send a fleet, accompanied by troops, on another operation against Yazoo City to prevent Confederates again from fortifying the town. Lieutenant Commander John G. Walker was in command of the *Baron de Kalb*. Major General Francis J. Herron was ordered to accompany him; his troops were not longer required to be at Port Hudson, which had recently capitulated.[38]

The *Baron de Kalb*, the side-wheel gunboats *New National*, *Kenwood*, and "stern-wheel" gunboat *Signal* were dispatched, under the command of Lieutenant Commander Walker, along with seven army transports bringing Herron's 5,000 infantrymen. The vessels proceeded up the Yazoo and encountered heavy artillery fire from the town. The *Baron de Kalb* returned fire sufficient to ascertain the strength of the water defenses, then dropped down the river to report to General Herron, who then landed his men and the army and navy then resumed a combined attack. As they were doing so, the Confederates began evacuating their works. Walker immediately steamed up and shelled the fortifications to prevent the withdrawal of the heavy guns. The Confederate regiment defending the town fled, leaving everything intact except for five Mississippi passenger river boats, which were all set on fire.[39]

Around 7:30 P.M., the *de Kalb*, leading the vessels, struck a torpedo while abreast of the navy yard at the lower end of town. As she was "moving slowly along" she hit an unseen torpedo, which sank her. As she was settling, another exploded beneath her stern. She was on the bottom in fifteen minutes. There were no serious injuries among her crew. The *de Kalb*'s guns were later salvaged and the hull was dragged into deeper water until the Federal forces could return and attempt raising her.[40] Once the *Baron de Kalb*'s other valuables had been recovered, and she being impossible to raise, she was blown up under water.[41]

The "unseen" torpedoes were of the Fretwell-Singer type, which along with the Rains keg torpedoes, were the types most frequently used by the Confederates. Admiral Porter later reported that "the usual lookout was kept for torpedoes, but this is some new invention of the enemy, which we will guard against hereafter."[42]

Captain Isaac N. Brown "late lieutenant in the U.S. Navy" led the Confederate withdrawal, set fire and sank fourteen other vessels, among them nine large ones whose machinery was intended to be moved to Selma, Alabama, for use in gunboats.[43]

It was later ascertained that seventeen torpedoes had been planted in the river. The river had risen two-three feet during the night, which allowed the other, shallower draft vessels, to pass over them before the *Baron de Kalb* hit one. Herron's forces captured 260 of the rear guard. Six heavy guns and the Confederate gunboat *St. Mary* fell into Union forces' hands. Punishing the citizens of Yazoo City for failing to warn the navy of the existence of the torpedoes, General Herron seized 3,000 bales of cotton "to pay for the gunboat that was lost through their treachery"[44]

The *de Kalb* was still in place in 1873 and last seen at low water in the 1950s.[45]

August 5, 1863: River gunboat *Commodore Barney* seriously damaged on the James River, Virginia

This side-wheel steamer, originally the ferry boat *Ethan Allen*, was purchased by the Navy in 1861. She weighed 512-tons and she was armed with seven guns.[46]

On August 4, 1863, Federal Major General John G. Foster, commander of the Department of Virginia and North Carolina, planned a reconnaissance up the James River. He was accompanied on the Army transport *John Farron* by Navy Captain Guert Gansevoort as well as Generals James S. Naglee and Robert B. Potter. Captain Gansevoort commanded the naval vessels of the reconnaissance: the steam frigate U.S.S. *Roanoke,* the monitor *Sangamon*, the *Commodore Barney*, and tug *Cohasset*, along with the army transport *John Farron.* The naval vessels were armed with artillery; the *Farron*, with a corps of sharpshooters. After spending the night anchored off Jamestown Island, the reconnaissance resumed their ascent of the James River with the army and navy officers now aboard the *Sangamon.* At 4:30 P.M., owing to the low water, General Foster and his staff, along with Captain Gansevoort, transferred to the *Commodore Barney*, and were followed by the *Cohasset.* Just beyond Cox's Landing, a torpedo, possibly two, exploded beneath the starboard bow of the *Barney*, "producing a lively concussion and washing the decks with the agitated water." Twenty men either jumped, or were swept, overboard; two were never recovered and presumed drowned. The "guard [was] perforated from hull on [the] starboard side forward for a distance of about 10 feet; [the] connecting steam whistle carried away, letting steam out of the boiler (repaired temporarily during the night)." With the steam pipe cut, the engine was partially disabled. The *Cohasset* took the *Barney* in tow and returned her to Dutch Gap at 7:30 P.M. During the next day the vessels slowly moved down the James River, coming under Confederate rifle fire from the shore.[47]

R.O. Crowley, an electrician of the Torpedo Division of the Confederate Navy, was present at the torpedo station. He recounted that the man responsible for making the electrical connection to fire the 1,000 pound galvanic torpedo lying on the river bottom, "lost his presence of mind and fired ... when the gunboat was at least twenty to thirty yards distant." The explosion threw a large column of water "to a considerable height," and the *Barney*, by her own momentum plunged, into the "great trough, and caught the downward rush of a wave on her forward deck"[48]

August 16, 1863: *Pawnee*'s launch sunk in Light House Inlet, South Carolina.

August 16, 1863: Gunboat *Pawnee* receives minor damage in Light House Inlet, South Carolina.

The *Pawnee* was a 1,289-ton twin-screw steamer built in 1860. She was armed with 20 guns.[49]

During the night of August 16, 1863, in Stono Inlet, South Carolina, Commander George B. Balch reported that at midnight a torpedo exploded under the *Pawnee*'s stern. Although it did not injure the ship, it destroyed the launch. Four hours later another torpedo exploded within 30 yards of the ship. The accompanying mortar schooner *C.P. Williams* captured two intact torpedoes. Balch later captured a boat with a platform for ten torpedoes.[50]

Balch later wrote more fully about the incident. Four torpedoes had exploded; and two were captured by the *Williams.* They contained 90 pounds of gunpowder. He found the torpedoes to be "ingenious and exceedingly simple" and sent drawings to Rear-Admiral John A. Dahlgren.[51]

These torpedoes had been designed by Confederate Captain M. Martin Gray and Stephen Elliot, who had released them 400 yards above the Federal fleet anchored in Light House Inlet on the night of August 11.[52]

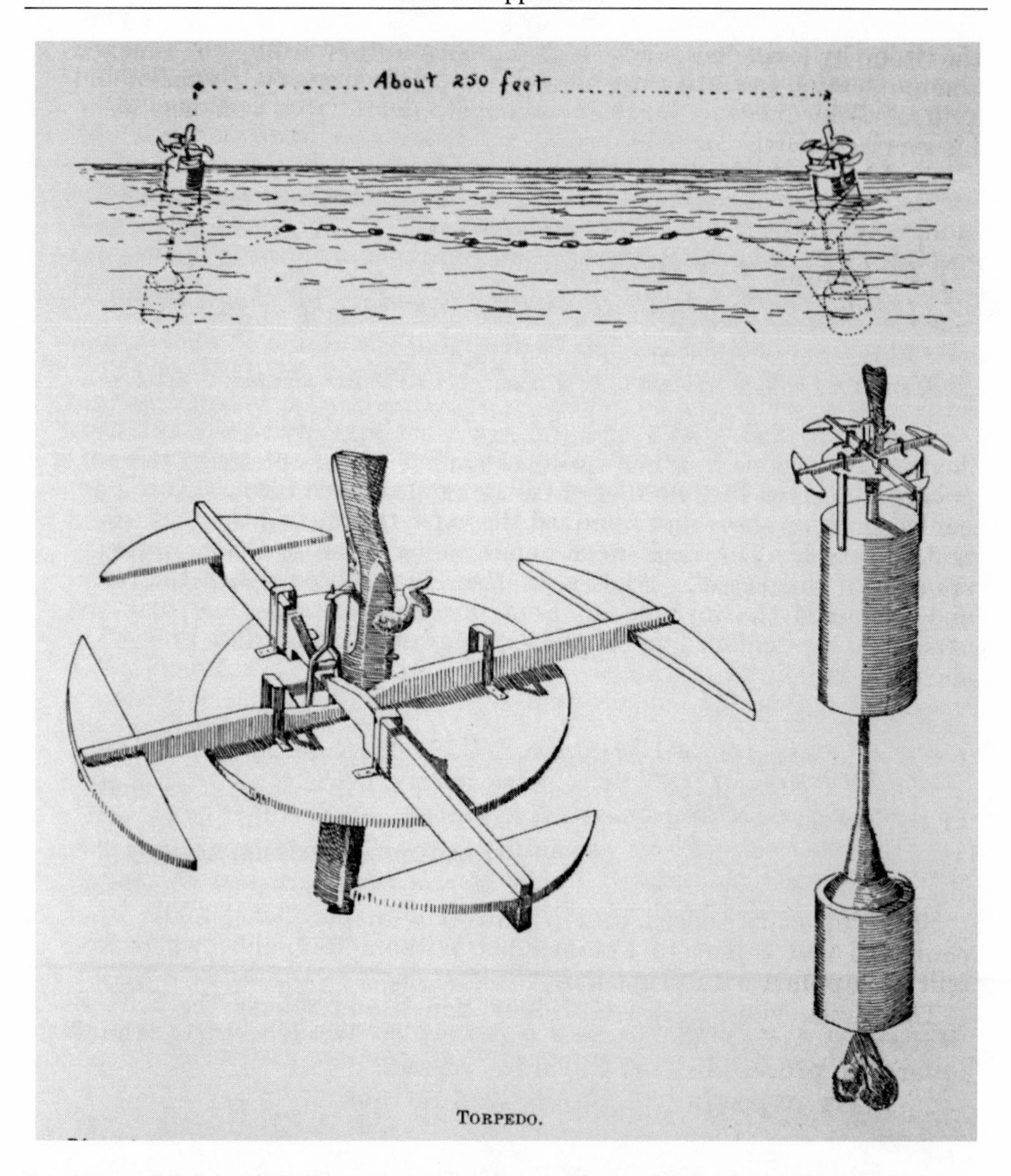

*Dimensions.—Whole length 10 feet; length magazine 18 inches; length of shaft 3 feet 6 inches; length of float 2 feet 4 inches; from top of float to trigger 10 inches; from trigger to end of musket stock 11 inches; diameter of float and magazine 15 inches; length of trigger arms 2 feet 1 inch; contained 60 pounds of powder.**

October 5, 1863: Ironclad *New Ironsides* receives serious damage in Charleston harbor, South Carolina.

This 3,486-ton ironclad was commissioned in 1862 and was armed with eighteen guns.[53]

On the evening of October 5, 1863, Confederate Lieutenant William T. Glassell, in command of a torpedo boat, along with Assistant Engineer James H. Tomb, Pilot J. Walker Cannon, and Seaman James Sullivan, left Charleston to attempt the destruction of the

**ORN*, Series 1, vol. 14, p. 447.

New Ironsides. The torpedo boat was painted "the most invisible color (bluish)." The torpedo was made of copper and contained 100 pounds of rifle powder. It had four "sensitive tubes of lead" and was attached to the "David" by a fourteen foot long shaft and was held at a downward angle which allowed it to pass six to seven feet below the surface. They passed through the Federal fleet in the darkness and approached the ironclad. About 9:15 P.M., and within 50 yards of the vessel, they were hailed by the watch from the ship. Receiving no reply, Union deck officer Charles W. Howard ordered "fire into her." To this, Glassell responded by discharging his double-barrel shot gun. Howard was seriously wounded by the Confederate's fire, and would die of his injuries five days later. Two minutes later the Confederate crew struck the ironclad with the spar torpedo "under the starboard quarter, about 15 feet from her sternpost, exploding our torpedo about 6½ feet under her bottom."[54]

The resulting explosion threw up a tremendous column of water which descended to extinguish the fires on the torpedo boat. Glassell and Sullivan were swept overboard; they were later captured. The pilot remained in the vessel; Tomb had jumped overboard, but quickly returned to the torpedo boat. Tomb was able to rebuild the fire in the engine and, after a delay, was able to get up enough steam to power the small vessel and escape, amid musket shots and the discharge of the 11-inch guns, to reach the Confederate docks in Charleston.[55]

As a prisoner, Lieutenant Glassell provided Rear Admiral Dahlgren with a sketch of the "David."[56]

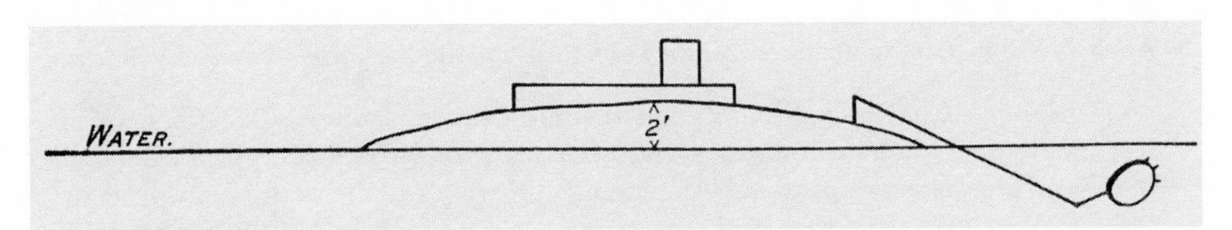

Dahlgren added that only two feet of hull projected above the water, below the water the hull extended for 50 feet in length. The torpedo was said to project about 10 feet beyond the front of the vessel. The fuse of the spar torpedo contained a vial of sulphuric acid; the torpedo contained 60 pounds of powder.[57]

Although initially thought to be unharmed by the explosion, the *New Ironsides* was later examined after her coal bunkers were nearly empty and found to be seriously damaged and in need of repairs. She was towed first to Port Royal and then to Philadelphia. Her hull plating and support beams had been markedly weakened and a large deck beam driven "on end." The *New Ironsides* would be out of service for a year.[58]

February 17, 1864: Screw sloop *Housatonic* sunk in Charleston Harbor, South Carolina.

This 1,240-ton screw steamer sloop-of-war was built on government contract and launched in November 1861; she was armed with thirteen guns.[59]

Admiral John Dahlgren initially reported that the *Housatonic*, which was on blockade off Charleston, South Carolina, was torpedoed by a rebel "David" and sunk on the night of February 17 at around 9:00 P.M. The deck officer reported that the attacking vessel was seen briefly before the attack. The *Housatonic* was struck on the starboard side, between the main and mizzen masts. She sank almost immediately; eight Union sailors were lost.[60]

Admiral Dahlgren observed: "I have attached more importance to the use of torpedoes than others have done, and believe them to constitute the most formidable of the difficulties in the way to Charleston. Their effect on the *Ironsides*, in October, and now on the *Housatonic*, sustains me in this idea."[61] Confederate prisoners, captured in a picket boat soon after the attack, "revealed" that the *Housatonic* was sunk by a "David" torpedo boat.[62]

Observation of the wreck three days later revealed "the after part of her spar deck appears to have been entirely blown off." Her cabin was completely demolished, as were all the bulkheads behind the mainmast.[63]

Engineer James Tomb, who was involved in the earlier attack in the "David" on the *New Ironsides*, observed a few months later that the *Hunley*, the Confederate submarine, which actually attacked the *Housatonic* from below the surface, rather than as a "David" on the surface, was a "veritable coffin." Like the "David" she employed a spar torpedo. She was powered by eight men and would make about three knots. "She was very slow in turning but would sink at a moment's notice and at times without it."[64] The *Hunley* never returned to Charleston. The trials and tribulations of the *Hunley* are well known and will not be recounted here. On August 8, 2000, the wreck was recovered, and may be seen today at the Warren Lasch Conservation Center in the former Charleston Navy Yard in Charleston, South Carolina.[65]

The hull and boilers of the *Housatonic* was located in 1980 and 1981; she was further investigated in 1996 and a few artifacts recovered in 1999.[66]

April 1, 1864: Army transport *Maple Leaf* sunk on the St. John's River, Florida.

The *Maple Leaf* was built in 1851 in Kingston, Ontario, and was a 398-ton wooden side-wheel steamer built to carry passengers, freight, and mail on both sides of Lake Ontario. In August, 1862, she was sold to a Boston firm and chartered to the United States government.[67]

The army vessels *General Hunter* and *Harriett A. Weed* were accompanying the *Maple Leaf* back from Palatka, Florida, on the St. Johns River.[68] The *Maple Leaf* was under the command of Captain W.H. Dale. Including the officers and crew, there were approximately 40 people on board.[69]

The *Maple Leaf* struck the torpedo twelve miles above Jacksonville at McIntosh's Point, opposite the mouth of Doctor's Lake, at 4:00 A.M., and sank in about seven minutes in eighteen feet of water. Two firemen and either two or three deckhands drowned. The baggage and camp equipment of the 112th, 169th NY and 13th Indiana regiments, along with the goods of two sutlers valued at $20,000, was on board; everything was lost. The regimental equipment had been loaded in Hilton Head, South Carolina, and was to have been unloaded in Jacksonville, but the vessel was then ordered to continue upstream to Palatka "on urgent business.[70]

The torpedo had been laid by Confederate Captain E. Pliny Bryan.[71] On the night of March 30, 1864, Bryan, a member of General P.G.T. Beauregard's staff, had placed "a number [12] of torpedoes in the channel" of the St. John's River above Jacksonville near Mandarin Point, rendering Federal communication with the garrison at Palatka "precarious." The torpedoes contained 75 pounds of gunpowder. Because of the heavy wind, Bryan could not board the wreck until the morning of April 2. He set 3 fires and she burned to the waterline. Bryan was assisted by members of the 1st Georgia Infantry and the 2nd Florida Light Artillery Battalion.[72]

In 1984, Dr. Keith V. Holland and a group of amateur historians located the site of the wreck at the bottom of the St. Johns River. In 1988 divers entered the hull, located beneath 20 feet of water and three to seven feet of mud, and removed nearly 3,000 items. The anaerobic mud and tannin rich water preserved the equipment of the three regiments and the sutlers. Some of the recovered artifacts are on display at the Jacksonville Museum of Science and History.[73]

April 9, 1864: Frigate *Minnesota* receives minor damage in Hampton Roads, Virginia.

This 3,307-ton screw steam wooden frigate was built in the Washington Navy yard and launched in 1855. She was armed with 40 guns.[74]

Lieutenant Commander John H. Upshur commanded the U.S.S. *Minnesota*. At 2:00 A.M. on April 9, 1864, a "dark object" was observed passing 200 yards away. Acting Ensign James Birtwistle was officer of the deck. Hailing the vessel, she answered "*Roanoke*." Birtwistle ordered the tug *Poppy* three times to go and see what the vessel was, but her steam was down and she didn't move. When the mysterious vessel moved rapidly toward the port side of the *Minnesota*, Birtwistle ordered the *Poppy* to "run that boat down."[75]

Before the *Poppy* could leave, the object, the Confederate torpedo boat *Squib* exploded her spar torpedo under the port side of the *Minnesota*. After the spar torpedo exploded the *Squib* backed off. Birtwistle jumped to a 9-inch gun and tried to train it on her, but the torpedo boat was inside of range. Sentries got off three rifle shots. Birtwistle was finally able to fire a 9-inch gun at the fleeing *Squib*. The tug *Poppy*, her steam still down, was unable to pursue the *Squib*. The torpedo boat successfully escaped in the direction of the Nansemond River. Several Federal vessels eventually went in search but could not find her.

Inspection of the *Minnesota* later revealed moderate interior damage. Some of her main support beams were split and her hull plates pushed inward by the force of the explosion; several of her guns were disabled.[76]

The Confederate spar torpedo contained 53 pounds of gunpowder. The spar torpedo used a sulphuric acid fuse.[77]

Fleet Captain John S. Barnes supplied the sketch of the torpedo boat shown below.[78]

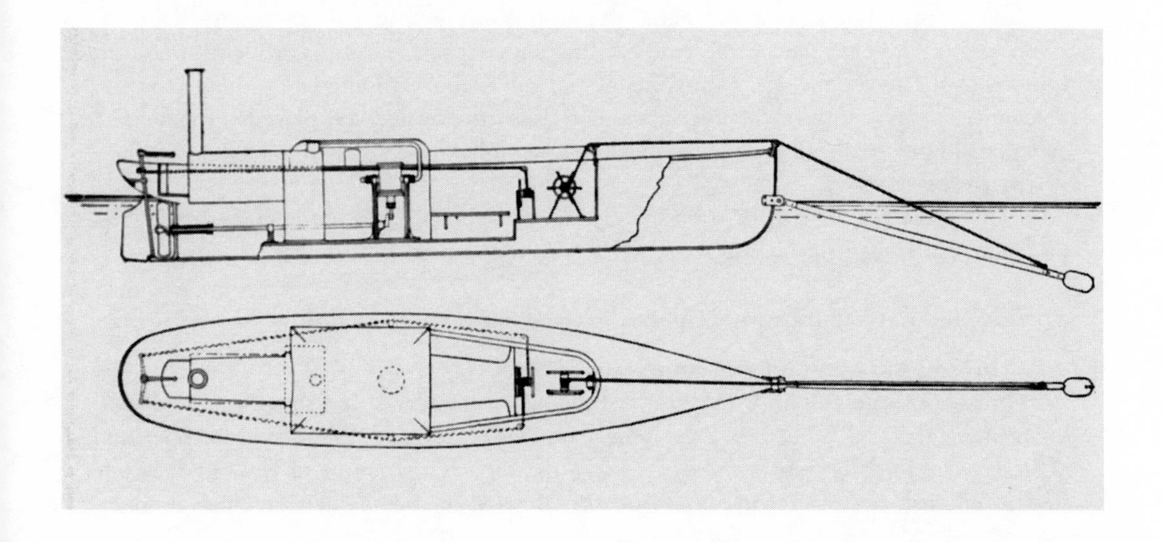

April 15, 1864: *Eastport*, armored river gunboat, sunk in the Red River, Louisiana.

Captured at Cerro Gordo, on the Tennessee River, Tennessee, on February 7, 1862, the *Eastport* was transferred to the Navy in January 1863. In Federal service she was a 700-ton ironclad steam gunboat, armed with eight guns.[79]

Her sinking was an episode which occurred near the middle of Major General Nathaniel Banks' Red River Campaign of March through May, 1864. On April 7, 1864, Rear Admiral David D. Porter's fleet of gunboats and transports conveyed Brigadier General T. Kilby Smith's division of the Sixteenth Army Corps on the Red River and its tributaries from Grand Ecore to Springfield Landing, arriving on April 10. There, the fleet found the channel of the river obstructed by a large sunken river steamboat, the *New Falls City*. By the time Smith's forces arrived, Banks' main force had been defeated in fighting at Sabine Crossroads and Pleasant Hill, and Banks sent word that he was falling back. Porter's gunboats were distributed among the troop transports and the vessels began to return down the river. Confederate infantry, now free of opposition, fired from the banks at the retreating flotilla. To compound this problem, the river was now falling and threatened to strand Porter's entire fleet. The river's "narrowness, and its high banks afforded the best possible opportunities for harassing attacks," and the bends of the river were so "that it was with greatest difficulty they were rounded by vessels." On April 15 the fleet was back at Grand Ecore.[80]

Lieutenant Commander S. Ledyard Phelps was in command of the *Eastport*, the largest of the ironclads accompanying the expedition. She had started down the river on the 14th and eight miles below Grand Ecore, she struck a torpedo and began to slowly fill with water. Within an hour of the explosion, the gunboat *Lexington* and the towboat *B* came along side to help keep the *Eastport* afloat. The damage was in the forward part of the vessel and only the bulkheads prevented the entire vessel from sinking.[81]

On April 16, Phelps lightened the ship by taking off her guns and heavy stores. The following day the steamer *Champion No. 5* arrived with two additional pumps which gradually began to "gain upon the water." On the 19th the steamer *New Champion* (*Champion No. 3*) arrived with her pumps and also began pumping out water. On April 21, fires were started in the furnaces of the *Eastport*, and one of the pumps was transferred to the *Eastport* from one of the *Champions*. All the while, carpenters from various vessels were at work trying to report the leaks.[82]

Despite the pumping and repairs, the river continued to fall and on April 26 the *Eastport* was hard aground on the sand and the network of sunken logs which formed the bottom of the river. She was approximately 1½ miles above Montgomery, Louisiana. All personnel and valuables were removed, and barrels of gunpowder were placed in the forward casemate, in the stern, and around her machinery—which, when ignited at 1:45 P.M., quickly destroyed the vessel.[83]

The wreck was located 1995 by the Corps of Engineers by magnetometer below 50–60 feet of sand, silt, and fill and some artifacts were removed.[84]

April 16, 1864: Army transport *General Hunter* sunk in the St. John's River, Florida.

This 350-ton transport was built as the *Jacob H. Vanderbilt* in 1863.

Brigadier General John P. Hatch, was Federal commander at Jacksonville, Florida. He directed that Colonel William B. Barton to oversee the evacuation of Palatka. A large portion of stores were initially transferred to Picolata, and then transshipped to Jacksonville. The steamer *Hunter*, on a return trip from Picolata, and having on board quartermaster's supplies, was destroyed by a torpedo at Mandarin Point, near the site of the wreck of the *Maple Leaf*. The *Hunter* was accompanied by the transport *Cosmopolitan*, and convoyed by the gunboat *Norwich*. The *Norwich* and *Cosmopolitan* passed single file over the torpedo without being aware of its presence. The *Hunter*, following in the wake of the *Cosmopolitan*, had difficulty staying on track due to heavy winds. She drifted toward the wreckage site of the *Maple Leaf* and struck the torpedo. The forward part of her hull fragmented and she sank immediately. The quartermaster on board drown.[85]

She was later raised and renamed the *General Sedgwick*.[86]

May 6, 1864: River gunboat *Commodore Jones* was sunk in the James River, Virginia.

The *Commodore Jones* began life as a ferry boat and was acquired by the Navy in May, 1863. She was a 542-ton side-wheel steamer armed with seven guns.[87]

As Major General Benjamin F. Butler's Army of the James began its ascent of the James River on May 5 to land on the Bermuda Hundred peninsula, seven side-wheel steamers, including the *Commodore Jones*, were sent to drag for Confederate torpedoes and intercept any fire-rafts which the Southerners might float down the river. The *Commodore Jones* and the side-wheel gunboat *Shokokon* would then anchor at the mouth of the Appomattox River between City Point and Bermuda Hundred, while their small boats would ascend the river in search of torpedoes. Only after the explosive devices had been cleared from the river could the ironclads ascend.[88]

The following day, May 6, the side-wheel gunboat *Mackinaw*, along with the *Commodore Jones* and side-wheel gunboat *Commodore Morris* moved to the James River. Commander John C. Beaumont of the *Mackinaw* had been ordered to remove torpedoes, whose location had been provided by a "contraband." Departing at 8:20 A.M., the vessels arrived at Jones Point, off Deep Bottom, in the James River, around noon. The three side-wheel steamers sent several small boats ahead, and then followed at a "safe distance." When the large ships were within 500 yards of the location indicated by the runaway slave, Beaumont ordered the three ships to drop anchor.[89]

After halting the *Jones* and *Morris*, Beaumont had the *Mackinaw* drop slightly downriver. Lieutenant Thomas Wade, commander of the *Commodore Jones*, ordered his vessel to slowly ascend the James in the wake of the smaller boats, which were still searching for torpedoes. The sailors in the rowboats safely passed over a large galvanic torpedo containing 1,750 pounds of gunpowder. The *Commodore Jones* slowly followed, unknowingly positioning herself close to the torpedo.

Confederate naval Lieutenant Hunter Davidson, knowing that "some of the many negroes [*sic*] prowling about" would tell the Federals about the torpedo station, had the wires and batteries shifted to the other side of the James River into a swampy area. The *Jones* was initially allowed to pass over the torpedo because the Confederates hoped to sink the flag ship or a monitor. When the *Jones*, after going ahead, began to drop back torpedo electrician Peter Smith closed the circuit.[90]

At 2:00 P.M. the torpedo exploded directly under the *Jones* "with terrible effect, causing her destruction instantly, absolutely blowing the vessel to splinters."[91]

An observer on a neighboring ship reported: "It seemed as if the bottom of the river was torn up and blown through the vessel itself. The *Jones* was lifted almost entirely clear of the water, and she burst in the air like an exploding firecracker. She was in small pieces when she struck the water again."[92]

Union Lieutenant Commander John S. Barnes later wrote that:

> Suddenly, and without any apparent cause, she appeared to be lifted bodily, her wheels rapidly revolving in mid-air; persons declared they could see the green sedge of the banks beneath her keel. Then through her shot to a height an immense fountain of foaming water, followed by a denser column thick with mud. She absolutely crumbled to pieces — dissolved as it were in mid-air, enveloped by the falling spray, mud, water, and smoke. When the turbulence excited by the explosion subsided, not a vestige of the huge hull remained in sight, except small fragments of her frame which came shooting to the surface.[93]

The *Commodore Morris* and *Mackinaw* immediately dispatched rescue boats. Prior to the explosion, the *Mackinaw* had sent a boat, under Acting Master's Mate Jeremiah F. Blanchard, to the left bank of the James near Four Mile Creek, to search for the galvanic torpedo station, its batteries and wires. Hearing the explosion, the men rushed back to their boat to help with the rescue effort. As Blanchard pulled survivors into his boat, he observed a man running on the opposite shore. The sailors fire several shots at him and he fell.[94]

After transferring the survivors, Blanchard and his party landed on the right bank of the James. He found two distinct batteries on shelves, both fully charged. After disconnecting them, he followed the wires 75 yards down to the river. He captured two Confederates, concealed in a small observation box in the ground. Blanchard reported that "the two wires running down the river bank were charged wires. They ran into this pit. The torpedo was exploded by applying one of the wires leading through the plug into the charged wires, thereby emitting a spark."[95]

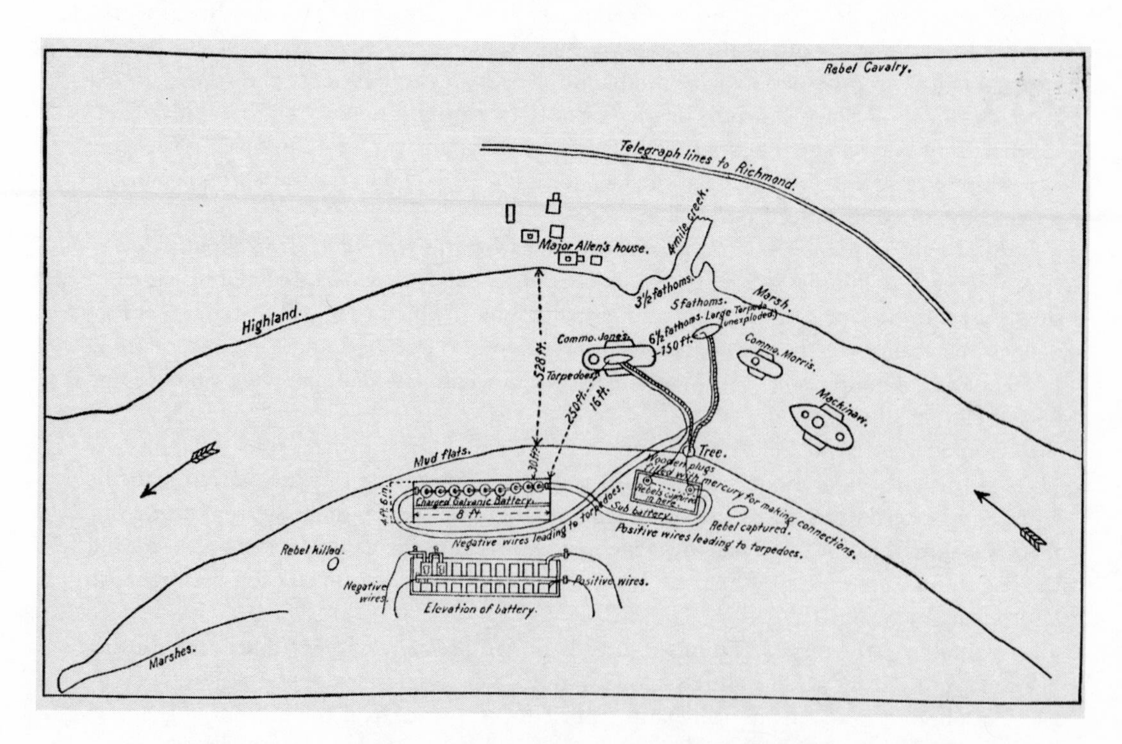

*Diagram of Galvanic Torpedo Station submitted by Assistant Engineer Jefferson Young.**

**ORN*, Series 1, vol. 10, p. 12.

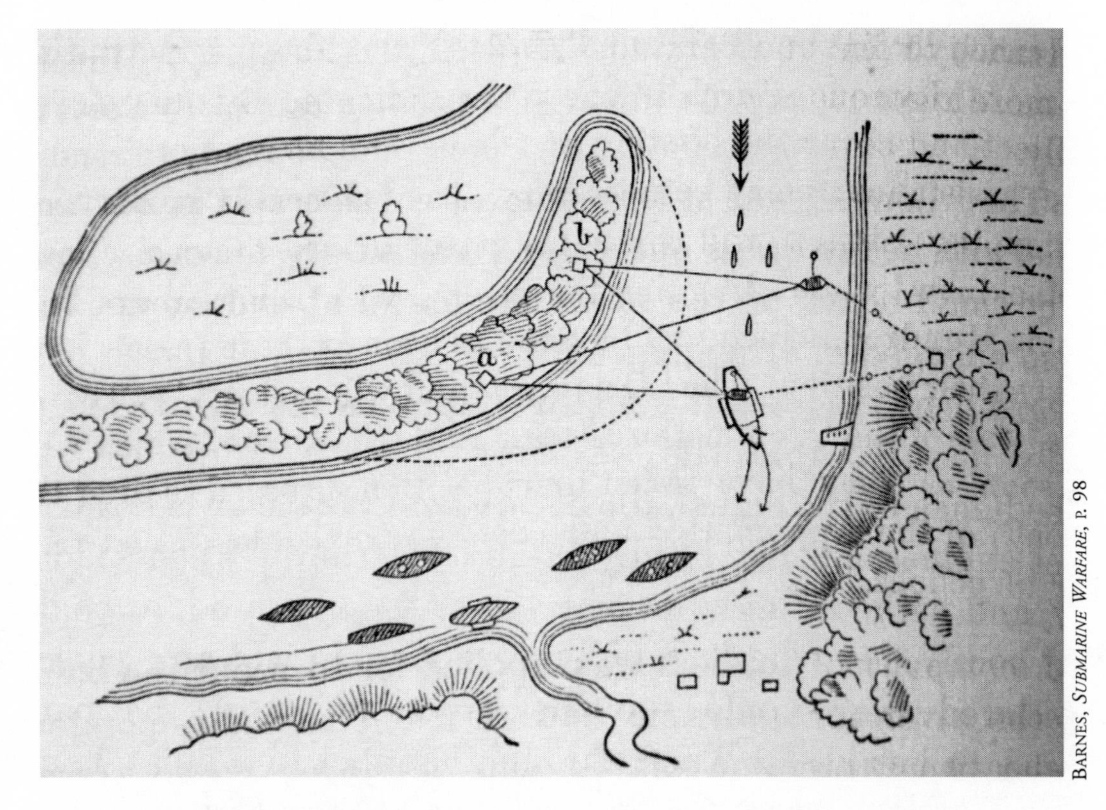

BARNES, *SUBMARINE WARFARE*, P. 98

Three observers triangulate on the Jones *to ascertain she is over the torpedo before exploding it.*

This was the first successful use of a galvanic torpedo to sink a vessel during the Civil War.

Approximately 40 men were killed in the explosion of the Commodore Jones.[96]

The final casualty list was completed days later; 69 men were dead, missing, or injured. Bodies were pulled from the river for six days.[97]

May 9, 1864: Army transport *Harriet A. Weed* sunk in the St. John's River, Florida.

This 210-ton side-wheel steamer was built in 1863.

Captain Gaskill was commander of the *Harriet A. Weed*, which had also served as a picket boat and was armed with two guns. The *Weed* was a little over two years old and had been used as a pilot boat off the St. John's bar. Around 8:00 A.M. on May 9, 1864, she left Jacksonville, Florida, towing a schooner, and in the St. John's River she hit two torpedoes twelve miles below Jacksonville, near the mouth of Cedar Creek, halfway to the St. John's bar. The *Weed* was totally destroyed, "literally blown to atoms." Five crewmen were killed; sixteen crewmen survived. In addition, the thirteen officers and 20 soldiers of the 3rd United States Colored Troops aboard were also saved. The schooner in tow also survived.[98]

July 19, 1864: Army transport *Alice Price* sunk in the St. John's River, Florida.

The 238-ton *Alice Price* was a side-wheel steamer.

In mid–July a Federal raiding party captured Holmes' saw mill on the Nassau River and dismantled its machinery, which was loaded aboard the *Alice Price.* The machinery was to be moved to Empire Mills near Jacksonville, where it was to be assembled and placed in service of the Federal troops. On her return upriver, the *Price* struck a torpedo near Beauclerc's Bluff, eight miles below Jacksonville and quickly sank. There were no casualties. The saw mill and the ship's machinery were salvaged.[99]

August 5, 1864: Monitor *Tecumseh* sunk in Mobile Bay, Alabama.

This 2,100-ton single-turret iron monitor was launched in September, 1863. She was armed with two 15-inch Dahlgren smooth bore guns. Commander Tunis A.M. Craven was in charge.[100]

Admiral David Farragut's naval attack into Mobile Bay began at 5:45 A.M. The wooden ships entered the main ship channel in the following order: The screw-sloop *Brooklyn* with the side-wheel steamer *Octorara* on her port side; screw-sloop *Hartford*, the flag ship, *with* the side-wheel steamer *Metacomet*; screw-sloop *Richmond* with the side-wheel steamer *Port Royal*; screw-sloop *Lackawanna* with the screw-sloop *Seminole*; screw-sloop *Monongahela* with the gunboat *Kennebec*; screw-sloop *Ossipee* with the *Itasca*; and screw-sloop *Oneida* with the screw-steamer *Galena.* The leading *Brooklyn* had "an ingenious arrangement for picking up torpedoes."[101]

The monitors *Tecumseh, Manhattan, Winnebago* and *Chickasaw* were positioned in a line on the starboard side of the fleet, with the *Tecumseh* in the lead. The *Tecumseh* and *Manhattan* were assigned to endeavor to destroy the Confederate ironclad ram *Tennessee*, which was in Mobile Bay.[102]

Confederate Lieutenant F.S. Barrett, Second Lieutenant in Charge of Torpedoes, noted that the Federal navy was "well informed as to the location of the torpedoes we had planted, as they kept well in on the east side of the channel where we had none, that part being left open by orders of the Chief of Engineer Department Lieutenant Colonel Viktor von Scheliha for the Confederate steamers to pass in and out.[103]

Fort Morgan opened fire at 7:06 A.M., and the shelling quickly became heavy on both sides. As the Federals steamed up the Main Ship Channel there was some difficulty and the *Hartford* passed on ahead of the *Brooklyn.* The monitor *Tecumseh* did not keep as far eastward as the other monitors, and at 7:40 A.M. she struck a torpedo and sank rapidly.[104]

The Confederates had planted 23 torpedoes in the main channel between to buoys and thirteen west of the west buoy and the line of piles opposite the water battery at Fort Morgan. The Federals has observed the passage of blockade runners and so had learned of the passageway in the channel. Lieutenant Barrett observed, "had their object not been to avoid them, they certainly would not have exposed themselves to the fire of Fort Morgan at such short range." He further noted: "The sinking of this monitor demonstrates the fact that if we had been allowed to plant torpedoes entirely across the channel, leaving no entrance for vessels to pass in and out, or even if we had extended our line 300 yards farther to the eastward very few, if any, of their vessels would have got through but many of them, no doubt, would have been sunk."[105]

Confederate Major General Dabney H. Maury, commander of the Department of the Gulf, thought the *Tecumseh* to have been sunk by a Fretwell-Singer torpedo. This was confirmed by Lieutenant Colonal Viktor von Scheliha. Von Scheliha had reported in July 1864 that 134 Singer's torpedoes and 46 of Brigadier General Rains keg torpedoes were aligned in three rows and in echelon in the shipping channel.[106]

The *Tecumseh* lost 93 of her crew, including Commander Tunis A.M. Craven; 21 men survived. The *Tecumseh* came to upside down in relatively shallow water of Mobile Bay. In 1873, she was sold for salvage. When family members of the lost crewmen learned that it was proposed to use explosives to break the wreck into salvageable pieces, they petitioned Congress to halt the work, which was quickly done. The remains of the *Tecumseh* are almost entirely buried in the soft mud and estimates of the cost to raise the vessel top eighty million dollars. In 1967 the anchor and dishes from the ship were brought to the surface.[107]

August 9, 1864: Supply ship *Lewis* and ammunition barges *General Meade* and *J.C. Kendrick*, sunk at City Point, Virginia.

The army supply ship *Lewis* and the two ammunition barges were sunk when a torpedo with a timer mechanism (an horological torpedo) exploded at the Union army's City Point, Virginia, warehouse complex on August 9, 1864.

Confederate Captain John Maxwell, of the Confederate Secret Service, left Richmond on July 26, 1864, for the James River, taking with him a horological torpedo to use against the vessels at the Union supply depot at City Point. He was accompanied by Mr. R.K. Dillard, who, being familiar with the area, served as a guide. Maxwell arrived in Isle of Wight County on August 2. Accompanied by Dillard, the two men passed through the picket lines on August 9, and, short of City Point, Maxwell then had Dillard wait for him. Continuing alone, Maxwell "approached cautiously" the last half mile to the wharf complex, with his "horological torpedo" in a box under his arm. In addition to the timer mechanism, it contained twelve pounds of gunpowder. When Maxwell was halted by a sentinel, he explained that he was delivering a package of candles. He was allowed to continue. After hailing a man on the barge *General Meade*, Maxwell set the timer mechanism and handed the device to him for delivery. Maxwell then rejoined Dillard.

The explosion destroyed the *Lewis* and the two munitions barges, as well as the adjacent warehouses. Maxwell, after reading newspaper accounts, reported that 58 were killed and 126 wounded, but he believed the number of casualties was much higher. He estimated the damage to be $4,000,000, although Federal authorities set the value at closer to $2,000,000.[108]

Lieutenant Morris Schaff, Ordnance Officer at the City Point depot, was present and recalled: "From the top of the bluff there lay before me a staggering scene, a mass of overthrown buildings, their timbers tangled into almost impenetrable heaps. In the water were wrecked and sunken barges, while out among the shipping — where were many vessels of all sizes and kinds — there was hurrying back and forth on the decks to weigh anchor, for all seemed to think that something more would happen."[109] Schaff reported that for months afterward it was thought the explosion was the fault of the Negro workmen who carelessly let a percussion shell fall.[110]

Lieutenant Colonel Horace Porter, General Grant's *aide de camp*, recalled "then there rained down ... a terrific shower of shells, bullets, boards, and fragments of timber. The general was surrounded by splinters and various kinds of ammunition, but fortunately was not touched by any of the missiles."[111] The reported casualties varied from 10 to 58 killed and 40 to 130 wounded.[112]

It was only after the war when the Confederate records were examined was the true cause of the explosion determined.[113]

November 27, 1864: Army transport *Greyhound* sunk in the James River, Virginia.

The *Greyhound* served at Major General Benjamin F. Butler's headquarters when afloat. She was a long, lean-looking craft, and the fastest steamer on the James River. On November 27, 1864, Admiral David D. Porter had ascended the river on his own ship, but received word that that Mr. Gustavus V. Fox, Assistant Secretary of the Navy, wished immediately to see him at Hampton Roads on important business, so Porter went down on Butler's *Greyhound*, which was faster than his own boat. "Every general of importance had a vessel for this purpose (carrying him from point to point), but the *Greyhound* was the gem of them all," Porter recalled. Five or six miles below Bermuda Hundred there was an explosion in the forward portion of the ship and smoke poured from the engine room. Within minutes the mid portion of the ship was in flames, and the upper saloon, occupied by Generals Butler and Robert C. Schenck, and Admiral Porter, filled with smoke. A small boat was lowered and the officers escaped. The *Greyhound* quickly became engulfed by flames from one end to the other. The officers were picked up by the steamer *Pioneer*, then transferred to the tug *Columbus* and taken to Fort Monroe. The *Webster* took off the rest of the crew and other passengers. Porter believed the cause of the explosion was "torpedoes ... throw[n] among the coal.... When the torpedo was thrown into the furnace with the coal, it soon burst, blowing the furnace-doors open and throwing the burning mass into the fire-room, when it communicated with the wood-work." He later opined, "In devices for blowing up vessels the Confederates were far ahead of us, putting Yankee ingenuity to shame."[114]

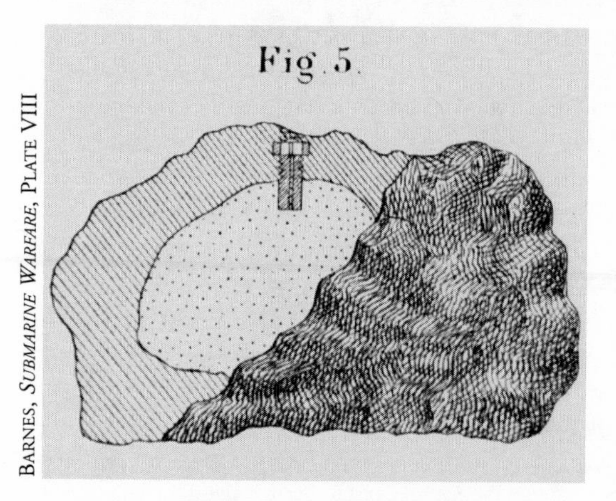

Barnes, *Submarine Warfare*, Plate VIII

The coal torpedo, a cast-iron block containing about 10 pounds of powder, was involved in the explosion of the *Greyhound*.[115]

Union naval Lieutenant John S. Barnes claimed that the Confederates had organized a body of men whose duty was to deposit these machines in coal-piles or barges, from which Federal vessels took their supplies. He reported that a large number of unaccountable explosions of Union vessels, principally transports, were attributed to these devices, the most notable of which was that of the "Greyhound."[116]

December 7, 1864: Tug *Narcissus* sunk in Mobile Bay, Alabama.

Purchased as the tug *Mary Cook* in September 1863, she was fitted and launched the following year. This 101-ton vessel was armed with 2 guns.[117]

The *Narcissus*, under Acting Ensign William G. Jones, she was on picket duty at Dog River Bar, Mobile Bay, on the night of December 7, 1864. She was anchored in eight feet of water "to the northward and westward of the fleet." Around 10:30 P.M. heavy squalls from the north and west threatened to ground the vessel, so Ensign Jones got her underway to the southeast about a mile, and anchored in nine feet of water. While paying out the anchor chain, the *Narcissus* struck a torpedo, which exploded and "lifting her nearly out of the water and breaking out a large hole in the starboard side, amidships, besides doing other damage, causing the vessel to sink in about fifteen minutes." Jones signaled his distress and in the morning the side-wheel steamer *Cowslip* came up and removed guns, ammunition, and small arms. Four men were seriously scalded when the steam pipe burst. Jones was later dismissed from the service after publishing a letter in a New Orleans newspaper about the circumstances of the loss of the *Narcissus*. The *Narcissus* was raised by the end of the year and, after being repaired in New Orleans, was back in service.[118]

December 9, 1864: Gunboat *Otsego* sunk on the Roanoke River, North Carolina. December 10, 1864: Picket boat *Bazely* sunk on the Roanoke River, North Carolina.

This *Otsego* was a 974-ton side-wheel steamer and was armed with ten guns. The 50-ton tug *Bazely*, originally the *Beta*, had been purchased by the Navy in June 1864 to serve at a picket boat.[119]

In December, 1864 General Grant ordered the Union army to move up the Roanoke River an additional 20 miles from Plymouth, North Carolina, and take Confederate Fort Branch, towering above the river at Rainbow Bluff. The operation was to coincide with Major General Benjamin Butler's operation against Fort Fisher to the southeast. The combined operation was initially to capture Fort Branch, on the Roanoke River, and then the infantry was to continue overland to Tarboro to destroy the railroad bridge across the Tar River and then fall back to the cover of the gunboats. The infantry brigade, under the command of Colonel Jones Frankle, consisted of the 85th New York Infantry; 27th Massachusetts Infantry; 9th New Jersey Infantry; 16th Connecticut Infantry; 176th Pennsylvania Infantry; and Battery A of the 3rd New York Heavy Artillery.[120]

The infantry left Plymouth on December 9 and the following morning skirmished with Confederates at Gardner's Bridge, and skirmished again that afternoon at Roster's Mills. That night the infantry occupied Williamston. In the morning gloom of the 12th, Frankle's men reached the rear of Fort Branch. As daylight dawned, they captured the confused Confederate commander, Colonel John Hinton, near his residence, and, posing as the Weldon Junior Reserves (then expected at Fort Branch), were able capture Confederates in the area. Moving to the crossroad at Butler's Bridge, Frankle encountered more of the enemy, and then returned to take Fort Branch. Finding it now fully garrisoned, and not having been supplied by the navy, Frankle elected to withdraw to Williamston. Along with the severe weather, "the failure of the navy, which had our extra ammunition, to connect and cooperate, and the lack of information from them, was the ground for withdrawal."[121]

Commander William H. Macomb directed the naval portion of the expedition, consisting of the side-wheel gunboats *Wyalusing*, *Otsego*, and *Valley City*, the tugs *Belle* and *Bazely*. Along with the infantry, the vessels departed Plymouth on December 9. Macomb had been assured the river had been cleared of torpedoes from Plymouth to Jamesville. Still, the boats had torpedo scrapers rigged on their bows extending 15 to 20 feet forward.[122]

Lieutenant Commander Henry N.T. Arnold was in command of the *Ostego*. As the fleet neared Jamesville, N.C., around 9:15 A.M. a torpedo exploded against the *Otsego*'s port side under the side-wheel and she began rapidly taking water. Three minutes later another explosion under the foremast caused the immediate sinking to the bottom in fifteen feet of water. There were no deaths.[123]

Since all vessels had been fitted with "torpedo catchers," the Federals reasoned that the torpedo which sank the *Otsego* was fired by "wires from the bank."[124] A guard boat to stand by the *Otsego* was requested, and it was planned to raise her, if possible.[125]

Squadron Commander Macomb left the *Otsego* where she was and continued up the

Roanoke River with his remaining vessels. The *Otsego* was left with four guns on deck behind breastworks of cotton bales, along with a good supply of ammunition. An awning had been rigged over the hurricane deck to shelter the crew, and the galley had been brought up from the main deck.[126]

The next day the tug *Bazely* went to the *Otsego*'s assistance, but as she came alongside, "a torpedo exploded beneath her, and blew her to fragments."[127]

During the next few days the Navy had learned that the Confederates had placed torpedoes in the Roanoke River, in the area of Sweetened Water Creek and until they could be removed, vessels should not ascend further up the river.[128] The gunboat *Valley City* was sent up from Plymouth to drag for torpedoes; the side-wheel gunboat *Ceres* was added to patrol the river.[129]

More than 30 torpedoes were then removed or exploded in the river. The remaining vessels could not ascend by steam, but had to use hawsers made fast to trees on the bank. They frequently jammed in the trees. Confederate infantry, now that Colonel Frankle had withdrawn, were free to harass the vessels from the high river banks.[130] After two weeks the Navy ended the mission.[131] The *Otsego*'s guns were removed and the wreck was destroyed[132]

January 15, 1865: Monitor *Patapsco* sunk off Charleston, South Carolina.

This 1,875-ton single turret monitor was built in 1862. She was armed with an 11-inch and a 15-inch Dahlgren gun.[133]

On January 15, 1865, the monitors *Patapsco* and *Lehigh* were the "picket monitor[s] of the night" and were engaged in covering the scout and picket boats searching for torpedoes and obstructions set by Confederates across the channel east of Fort Sumter.[134]

Lieutenant Commander Stephen P. Quackenbush was in command of the *Patapsco* when she hit a torpedo in the water between Forts Sumter and Moultrie around 8:15 P.M. She had been in the general vicinity of the "Lehigh buoy," marking when the monitor *Lehigh* had previously gone aground. The *Patapsco* had her torpedo fenders and netting deployed as usual. Three boats with drags had preceded her.[135]

The torpedo exploded on the port side 30 feet behind the bow. Although Quackenbush ordered the pumps started the forward end of the vessel was soon under water and it was clear that she could not be saved. Before the men below deck could be evacuated, the ironclad sank to the top of the turret. Quackenbush later stated that "at no time did I apprehend any danger whatever from torpedoes, as it was generally supposed that they were sunk above the line from Moultrie to Sumter." The monitor sank in 30 feet of water 800 yards from Ft. Sumter and 1,200 yards from Ft. Moultrie. Five officers and 38 men were saved; 62 officers and men drown.[136]

Initial Confederate reports erroneously claimed the *Patapsco* had been sunk by a torpedo boat.[137]

Confederate John A. Simon, Captain in Charge Torpedo Service, reported two days after the sinking of the *Patapsco* that he had been engaged for some ten days before in placing torpedoes in the location when the monitor was struck. "For some time past the enemy's picket-monitors have been in the habit of venturing much closer in the harbor than usual, and it has been my ambition to teach them a lesion, as well as our friends, upon the subject of torpedoes." He concluded: "One of these turreted monsters has met a fitting fate."[138]

Admiral John Dahlgren observed of the sinking and the problems which his blockading fleet outside of Charleston Harbor experienced: "No one who has not witnessed it can appreciate the harassing nature of the never-ceasing vigilance with which the monitor duty is sustained in this harbor." Indeed, the Confederates felt that Admiral Dahlgren was "more interested in protecting his ships from torpedo attack than in forcing an entrance into the harbor."[139]

Most of the wreck was removed in 1873.[140]

Feb. 17, 1865: Confederate flag-of-truce steamer *Schultz* sunk in the James River, Virginia.

The 164-ton side-wheel steamer *Schultz,* built in 1850, had been a "James River steamer and excursion boat" which had been acquired and modified by the Confederates to serve as a flag-of-truce boat for the exchange of furloughed prisoners of war.[141]

On February 17, with Captain Hill at the helm, she passed down the James River carrying 700 Union prisoners to be exchanged. The paroled Confederate prisoners who were to be exchanged and returned to Richmond were not awaiting the *Schultz,* and so she returned with only a few members of the "Richmond ambulance committee" and the usual crew. A few miles above Cox's Landing, she struck a loose Confederate torpedo and quickly sank. Two Confederate guards and two firemen were lost.[142]

Confederate naval Captain John K. Mitchell reported two days later that "the steamer *Schultz* has been blown up by a torpedo just below the bluff at Bishop's 60 yards from the south bank, probably one of those placed by Lieutenant [Beverly] Kennon and drifted from its original position."[143]

On February 22, Judge Robert O. Ould, Confederate Commissioner for the Exchange of Prisoners, wrote Union Major General Edward O.C. Ord, asking that a truce be held where the *Schulz* had sunk, to investigate the cause of the accident. General Grant wrote Ord, not objecting to the truce, but adding that he did not want the machinery of the *Schultz* being gathered by the Confederates to again be used against the United States forces, "unless it is shown that the accident occurred from a torpedo put in the water by us there is no claim upon us for its recovery."[144]

February 20, 1865: Gunboat *Shawmut*'s launch sunk in Cape Fear River, North Carolina.

On the morning of February 18, 1865, the following vessels went into action against Fort Anderson: monitor *Montauk*, gunboats *Huron* and *Shawmut*, and side-wheel steamers *Mackinaw*, *Lenapee*, *Maratanza*, *Osceola*, *Pontoosuc*, and *Pawtuxet*. At mid-day they engaged Fort Anderson and the bombardment continued until late in the afternoon. Early the following day the Confederates abandoned the fort and the *Shawmut* anchored five miles above Fort Anderson.

On February 20, having passed above Fort Anderson, the side-wheel gunboat *Eolus* moved upriver toward the Confederate batteries. During the late afternoon she was joined by the entire fleet, which shelled Confederate Fort Strong (composed of Batteries Meanes, Campbell, Lee, and Davis), on the left bank of the Cape Fear River, below Wilmington near when the Brunswick River joins the Cape Fear River.[145]

Admiral Porter learned from a contraband that the Confederates were planning "to let a hundred torpedoes drift down upon us at night and blow us all to pieces!" Although he had little confidence in the story, he ordered a double line of nets placed across the river to intercept the "visitors." Most were intercepted, but a few passed through.[146]

Around 10:00 P.M. the *Shawmut*'s forward lookout saw something floating down in the ebb tide. Lieutenant Commander John G. Walker, commanding officer of the *Shawmut*, called all hands called to quarters. He sent Acting Ensign William B. Trufant away in the first cutter to examine the objects, there being a large number in sight. He sent away John A. Davis in the gig. Davis later returned with five torpedoes, which he hauled to the *Shawmut* and made fast to the stern. Mr. Trufant tried to sink one of the torpedoes by firing at it with his pistol. Unfortunately, it exploded, killing two men and wounding Traufant, and sinking the launch.[147]

February 21, 1865: Gunboat *Osceola* moderately damaged in Cape Fear River, North Carolina.

This 974-ton side-wheel double-ender gunboat was delivered to the navy in January, 1864. She was armed with eight guns.[148]

Commander John M.B. Clitz captained the *Osceola*. After the evacuation of Fort Anderson on February 19, he, along with the other vessels of the fleet, continued the ascent of the Cape Fear River, supporting the Federal troops advancing on both sides of the river—Major General Jacob D. Cox on the right bank and Major General Alfred H. Terry on the left bank.[149] After sounding and buoying out the middle ground at Big Island, the gunboats opened fire on Fort Strong, the work commanding the principal obstructions, where the rebels had also sunk a large steamer, the *North Easter*.

On the night of the 20th, Commander Clitz also reported that the Confederates sent down 200 floating torpedoes, but a strong force of picket boats sank most with musketry. Some of the vessels picked up the torpedoes with their torpedo nets. Despite all these precautions, one got in the wheel of the *Osceola* and blew her wheelhouse to pieces, and knocked down her bulkheads inboard, but did not damage the hull.[150]

March 1, 1865: Gunboat *Harvest Moon* sunk in Winyah Bay, South Carolina.

This 546-ton side-wheel steamer was purchased by the Navy in 1863; she served as Admiral John A. Dahlgren's flag ship. She was armed with six guns.[151]

Acting Master John K. Crosby captained the ship. The *Harvest Moon* had been near Georgetown, South Carolina, until April 30, and then dropped down 2 or 3 miles to captured Confederate Battery White.[152] At 7:15 A.M. she got underway and proceeded down the river through Marsh Channel, accompanied by the tug *Clover*. At 7:45 A.M., when about 3 miles from Battery White, she hit a torpedo. It blew a hole through the starboard quarter, tearing away the main deck over it; the ship sank in five minutes in fifteen feet of water. The tug *Clover* immediately came to her assistance. Admiral Dahlgren and his staff went on board *Clover*; the ship's officers remained on board to save everything possible. A gig in charge of Acting Ensign D. B. Arey went to the gunboat *Pawnee* for assistance. In addition, three boats went up the river to drag for torpedoes. John Hazzard, wardroom steward, was killed in the explosion.[153]

The *Harvest Moon* remains in Winyah Bay, her stack sticking above the surface. The state of South Carolina is considering plans to raise the *Harvest Moon* as a Civil War Monument.[154]

March 4, 1865: Army transport *Thorn* sunk in the Cape Fear River, North Carolina.

The *Thorn*, a 403-ton army transport vessel, had been employed since Brigadier General Ambrose E. Burnside's campaign in eastern North Carolina in 1862. In March 1864 she was transporting goods to and from Federally occupied Wilmington from ships too heavy to pass the bar at the mouth of the Cape Fear River. After the fall of Fort Anderson of February 18, Admiral Porter had the river "thoroughly dragged" for torpedoes. The *Thorn* was heavily loaded with exchanged Union prisoners who had been held by the Confederates in Georgia and South Carolina. The gunboat *Eolus* had passed down the river half an hour before, and other vessels had been passing up and down continually during the day. The explosion, just below Fort Anderson, occurred under her bow, and she sank within two minutes. The number of casualties is not known. There were suspicions that the torpedo had been planted "by a prowling band of rebels."[155]

March 6, 1865: Tug *Jonquil* seriously damaged in the Ashley River, Charleston, South Carolina.

This 90-ton light gunboat, originally the *J.K. Kirkman*, was purchased by the Navy in October, 1863. She was armed with two guns.[156]

Acting Ensign Charles H. Ranson was in command of the *Jonquil*.[157] On March 5, she was dragging for torpedoes, accompanied by two boats from the five gun steamer U.S.S. *Home*, in the mouth of the Ashley River. He soon hooked on to something which the boats could not stir. After maneuvering and hauling, it came slowly, but owing to the great weight it broke. Those portions hauled aboard proved to be a framework of pine logs, spiked together with heavy iron spikes. It was 20 to 30 feet square; Ranson reported that "on each end of the frame was placed a torpedo, made of iron, and conical, having on their bottom four flanges for bolts and nuts, which were riveted to the logs. The torpedoes were capable of holding from 30 to 40 pounds of powder each, having a percussion fuse, to be ignited by sulphuric acid in a glass vial." The frame had been filled with stones and at low tide, was about two feet below the surface. At high tide it was about eight feet below the surface. Ranson continued dragging for the rest of the frame torpedo, and eventually pulled it to pieces.

The following day, the 6th, Ranson continued his dragging operation. He quickly hooked another frame. It was about 100 yards above the previous one, a little towards the right. The frame became separated and he successfully removed three of the four iron torpedoes fastened to the ends of the logs. He secured the final log in such a way that the torpedo portion remained beneath the surface of the water. As he approached the shallower water of the shore to destroy the torpedoes, the one underwater struck the bottom and exploded directly under the mid-portion of the *Jonquil*. "Every movable thing was thrown down, doors shattered, windows all broken, and all light work started. The howitzer forward was upset and three beams were badly sprung. The steam gauge and condenser were broken and nearly all the lighter machinery was disabled. The hull of the vessel, however, I found on examination, was not materially damaged."

On the 7th, with the *Jonquil* repaired, Ranson continued his work of dragging for torpedoes. He found yet a third frame torpedo, which he successfully and uneventfully removed from the river. He noted "From the positions they were placed, the three frames formed a triangle; thus, had a vessel escaped the first set she would very likely have fouled one of the three others."[158]

March 12, 1865: Tug *Althea* sunk on the Blakely River, Alabama.

In January 1865 Lieutenant General Grant ordered Major General Edward R.S. Canby to move against Mobile, Alabama. The West Gulf Blockade Squadron was now under the command of Admiral Henry K. Thatcher. Canby believed an advance up the eastern side of Mobile Bay, with the capture of Spanish Fort and Fort Blakely, would force the surrender of Mobile. The March and April 1865 losses in Mobile Bay and the Blakely River are associated with naval support in Canby's campaign.[159]

The *Althea*, a 72-ton tug was purchased by the Navy in December 1863. She was armed with one gun.[160]

Ensign Frederic A.G. Bacon, *Althea*'s commander, reported that he was ordered to drag the channel for torpedoes with a chain attached to spars laid across the stern. Pilot Jesse Dinton accompanied Bacon to instruct him where to go. When the Althea was abreast of Confederate Battery Huger, the chain fouled on an old wreck. Unable to pull the chain out, he released it into the water. As he attempted to return to the side-wheel gunboat *Octorara*, he ran into a torpedo, which exploded near the after part of the pilot house, a little to the starboard. The *Althea* sank immediately in ten to twelve feet of water. Two men were killed, and Ensign Bacon, along with two other crewmen, was badly injured.[161]

The *Althea* was raised in November 1865.[162]

March 17, 1865: Coast survey steamer *Bibb* was slightly damaged in Charleston Harbor, South Carolina.

The 409-ton coastal survey steamer *Bibb* was built in 1845; she was converted into a side-wheel vessel in 1861.[163] Captain Charles O. Boutelle, commander of the *Bibb*, was marking the course of vessels between Forts Sumter and Moultrie when, at 5:25 P.M., she struck a torpedo, which exploded under the port bow "midway between the port guard and the fore channels." The column of water thrown up nearly filled one of the ship's cutters. Portions of the port side frame were loosened, and some of the steam pipes were broken on the sides. The torpedo was struck "directly in the track over which many vessels have passed" and Boutelle concluded that it must have been ten feet below the surface, which was the depth of the *Bibb*'s hull.[164]

March 28, 1865: Monitor *Milwaukee* sunk in Blakely River, Alabama.

This 970-ton double turret ironclad was delivered to the Navy in August 1864. She was armed with four guns.[165]

The *Milwaukee*, under the command of Lieutenant Commander James H. Gillis, was proceeding up the Blakely River, accompanied by the monitor U.S.S. *Winnebago*. They were approaching within 1½ miles of Spanish Fort for the purpose of shelling a Confederate transport thought to be supplying the fort. The enemy steamer had retreated up the river, and the *Milwaukee* was returning to her former position, within 200 yards of the anchored ironclad *Kickapoo*. Union boats were actively sweeping for torpedoes, and in the spot where the *Winnebago* had just turned not ten minutes before, the *Milwaukee* struck a torpedo on the port side. The stern of the vessel sank in about three minutes; the forward compartments did not fill until an hour later, allowing the entire crew to be safely taken off by the *Kickapoo*.[166]

March 29, 1865: Monitor *Osage* sunk in Blakely River, Alabama.

This 523-ton single turret monitor was launched in 1863; she was armed with two 11-inch Dahlgrens.[167]

Lieutenant Commander William M. Gable, commanding officer of the *Osage*, reported that at 2:00 P.M. he was anchored inside the bar of the Blakely River, along with ironclads *Kickapoo*, *Winnebago*, and *Chickasaw*, and gunboat *Ottotara*. A strong breeze was blowing eastward, and the *Osage* began drifting towards the *Winnebago*.

As she moved ahead of the *Winnebago*, off her starboard bow, and halted in twelve feet of water, a torpedo exploded beneath the bow and the *Osage* immediately began sinking. The area of anchorage of the *Osage* had been "thoroughly dragged by boats" and Gable believed that the "torpedo by which she was sunk was submerged and drifting." Four men were killed and 8 men were injured.[168]

She was raised and sold for scrap in November 1867.[169]

April 1, 1865: Armored river gunboat *Rodolph* sunk in the Blakely River, Alabama.

This 217-ton side-wheel steamer was purchased by the Navy in December 1863. She was a "tin clad" gunboat and armed with twelve guns. She was also known as *Tin Clad No. 48*.[170]

Acting Ensign James F. Thomson was acting commander of the *Rodolph* on April 1; her regular chief, Acting Master N. Mayo Dyer, was aboard the side-wheel gunboat *Metacomet*. At 1:00 P.M. the *Rodolph* was ordered to take along side a barge containing pumps for raising the *Milwaukee*. She, and the barge, crossed the bar of the Blakely River and moved toward the *Milwaukee*. Around 2:40 P.M. she was between the ironclads *Chickasaw* and *Winnebago*, when a torpedo exploded under the starboard bow, blasting a ten foot hole in the hull and causing the rapid sinking of the *Rodolph* in twelve feet of water. One man was killed, eleven wounded, and three men were missing as a result of the explosion. Her guns and valuables were later salvaged.[171]

April 13, 1865: Tug *Ida* sunk in the Blakely River, Alabama.

This 104-ton steam tug was purchased by the Navy in April 1863. She was armed with one gun.[172]

Acting Ensign Franklin Ellms was in command of the *Ida* on the morning of April 13, 1865. He was ordered first to the monitor *Milwaukee*, then to proceed to the side-wheel gunboat *Genesee*, then lying two miles below the obstructions on the Blakely River. The *Ida* had sailed about two-thirds of the distance when she struck a torpedo, crushing her starboard side, bursting her boilers, and tearing up the decks. She sank in the main ship channel in a few minutes in ten feet of water in the mid-channel, near Choctaw Pass. One man was killed, three wounded, and one man was missing as a result of the explosion and sinking.[173]

She was raised and sold in September 1865.[174]

April 14, 1865: Gunboat *Sciota* sunk in Blakely River, Alabama.

This 507-ton screw steamer gunboat was delivered to the Navy in November 1861. She was armed with five guns.[175]

Volunteer Lieutenant James W. Magune of the *Sciota* had just delivered a coal barge from the brig *American Union*, and had taken working parties to the gunboats *Itasca*, *Sebago*, and *Genesee*. The *Sciota* was moving southeasterly to deliver another work party to the side-wheel gunboat *Elk* when she hit a torpedo. The blast broke the beams of the spar deck, "tearing open the waterways, ripping off starboard forechannels, and breaking fore-topmast." The *Scotia* immediately sank in the shallow water to the deck line. Three men were killed in the explosion.[176]

She was raised by Union forces in July 1865.[177]

April 14, 1865: Ironclad *Cincinnati*'s launch sunk in the Blakely River, Alabama.

The launch was part of the continuous search for Confederate torpedoes in Mobile Bay and the tributary rivers. Her crew had caught a torpedo and was slowly pulling it toward the launch when a rope broke and torpedo hit the stern of the boat and exploded. Three men were killed.[178]

April 25, 1865: Army transport *R.B. Hamilton* sunk in Mobile Bay, Alabama.

The *R.B. Hamilton* was a 175-ton stern-wheel steamer built in 1858. She was traveling from New Orleans, Louisiana, to Mobile, Alabama. She was carrying three companies of the 3rd Michigan Cavalry. In the Lower Gap into Mobile Bay she struck a floating torpedo and was "completely wrecked." Fifteen persons were killed or wounded; six "firemen" were among the killed.[179]

The following five vessels were reported by other authors as sunk or damaged but I have been unable to confirm the incidences.[180]

September ——, 1863: Army Transport *John Farron*, reportedly seriously damaged by a galvanic torpedo on the James River, Virginia.[181]

The *John Farron* was a 250-ton side-wheel steamer built in 1856.[182]

March 6, 1864: Gunboat *Memphis* receives no significant damage in the North Edisto River, South Carolina.

This 791-ton screw steamer was originally a Confederate blockade runner when captured by the U.S.S. *Magnolia* on July 31, 1862. She was converted to a gunboat later that year and she was armed with seven guns.[183]

Chief Engineer James Tomb was in command of the cigar torpedo steamer which struck the *Memphis* in the North Edisto River, but he failed to destroy the gunboat because the spar torpedo did not explode. The torpedo employed had been "exposed" to "every vicissitude of weather and climate" for the past six months, and Confederate Chief of Engineers, Captain Francis D. Lee, had told him to replace it with a new, tested torpedo, which Tomb neglected to do.[184]

The attack had originally been planned for March 4, but the David's pumps failed, as they did again the following night when another attack was planned. After making repairs, Tomb set out around 12:30 A.M. on March 6 in a third attempt to attack the *Memphis*. Tomb had protected the open hatch with steel plate, and small arms fire from the *Memphis* had no effect.[185]

Acting Master Robert O. Patterson, commander of the *Memphis*, reported that at 1:00

A.M. on March 6, 1864, he observed a lead gray torpedo boat descending the North Edisto River rapidly and approaching the port quarter of the *Memphis*. The crew was beat to quarters and the chain was slipped, but in an instant the torpedo boat was beneath the port quarter. The armed watch was able to concentrate rifle and revolver fire on her hatchway, and the boat stopped and dropped astern twelve feet, then darted ahead again toward the *Memphis*. Patterson had ordered the vessel go ahead, and the torpedo boat was damaged by the ship's propeller.

Tomb was able to strike the *Memphis* about eight feet below the surface but the spar torpedo failed to explode. He came in for a second strike, but only got off a glancing blow and, again, the torpedo failed to explode.[186]

The smaller boat then seemed to drift off and moved back up the river. When a light became visible on the torpedo boat, the *Memphis* fired on her with a 12-pounder rifle, but the torpedo boat disappeared in the darkness. The *Memphis* dispatched an armed boat to search for her, but was unsuccessful in finding her.[187]

Returning up the North Edisto River to Church Flats, Tomb examined the torpedo. He saw that the cap covering the primer was smashed flat and the tube containing sulfuric acid broken; this was from the first strike. The second glancing strike resulted in a slightly indented primer cap with the underlying tube of acid intact. Neither fuse had detonated the 95 pounds of gunpowder in the spar torpedo.[188]

April 14, 1865: Screw steamer *Rose* reportedly sunk in Mobile, Bay, Alabama.[189]

This 2-gun, 96-ton armed tug boat was purchased under the name *Ai Fitch* in 1863.[190]

April 15–24, 1864: Troop transport *St. Mary's* reportedly sunk in Mobile Bay, Alabama.[191]

March 19, 1865: Navy transport steamer *Massachusetts* not damaged in Charleston Harbor, South Carolina.

This 1,155-ton screw steamer supply ship was purchased by the Navy in 1861. She was armed with seven guns.[192]

Acting Volunteer Lieutenant Commander William H. West reported that on March 19 that the *Bibb* was approximately 50 yards from the buoy marking the *Patapsco* wreck when the *Massachusetts* struck a torpedo which failed to explode. "The keel must have torn it from its moorings, for it struck the ship heavily under the starboard quarter and came up to the surface from under the propeller cut in two. I endeavored to pick it up, but before the boat reached it had sunk."[193]

*Editor's Appendix 2: Examples of Extant Confederate Torpedoes**

Subterra Torpedo. *United States Military Academy Museum Accession: 3673.*

This Confederate torpedo was submitted to the museum by Major General Absalom Baird. It was part of a row of torpedoes stretching across the Louisville Road in front of the Confederate earthworks west of Savannah, Georgia. This is an unusual design for a subterra torpedo; I think the device is too small to have originally been designed as a spar torpedo.

This specimen is 15 inches long and 7 inches in diameter (Image: Herbert M. Schiller).

*The source of the image is listed in the caption. Two photographs supplied by the United States Military Academy and ten engravings are identified as to their origin. Additional examples of torpedo specimens can be found in Jack Bell, *Civil War Heavy Explosive Ordnance* (Denton: University of North Texas Press, 2003), pp. 474–496, and Kochen and Wideman's *Torpedoes, Another Look at Infernal Machines.*

Subterra Torpedo. *This Confederate 10-inch mortar shell has been converted into a subterra shell. The fuse is in the upper center; the loading port is in the lower center. The fuse is covered by the safety cap, which would be removed after the torpedo had been placed but before it was covered. The device would be placed in the ground, protected by a tin shield or by a board, and covered with a small amount of soil. (See Rains* Torpedo Book, *Plate 13).*

This specimen is 9.87 inches in diameter.

This specimen was recovered at Fort Blakely, Alabama, part of the outer Mobile defenses. The fuse does not have are large hexagonal nut for tightening. The shell was found barely exposed in a logging road, having been repeatedly run over by trucks (Image: Atlanta History Center).

Subterra Torpedo. *This Confederate 24-pounder shell has been converted into a subterra shell. It was found at Blakely, Alabama. The safety cap has been removed. The cover for the contact fuse is dented. The loading port is on the side of the shell.*

It is 5¾ inches in diameter; its height, with the fuse, is 7½ inches (Image: Jack Bell).

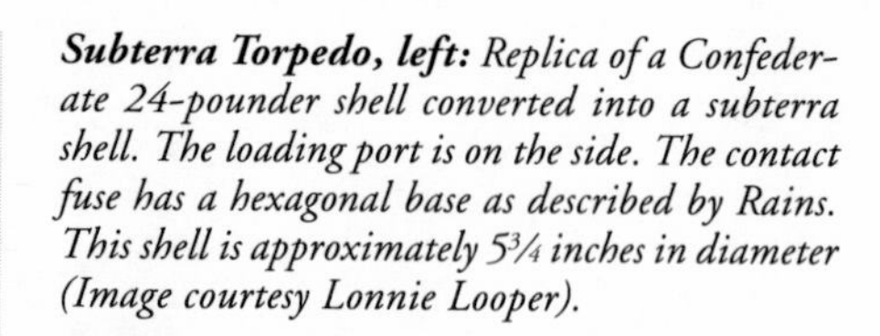

Subterra Torpedo, left: *Replica of a Confederate 24-pounder shell converted into a subterra shell. The loading port is on the side. The contact fuse has a hexagonal base as described by Rains. This shell is approximately 5¾ inches in diameter (Image courtesy Lonnie Looper).*

Fretwell — Singer Torpedo, below and facing page: *United States Military Academy Museum Accession: 3674.*

This torpedo was presented to the museum by Lieutenant Guy V. Henry when he was participating in operations around Petersburg and Richmond in May 1864 to April 1865. This specimen was presumably manufactured at the Richmond Torpedo Factory, to be used in the James River.

This is a Fretwell-Singer percussion torpedo. It was invented by E.C. Singer, nephew of Issac Singer, the sewing machine inventor, and improved upon by the Texas physician John Richard Fretwell.

The tapered tin case is 13½ inches long and 12¼ inches wide. Powder was in the bottom of the container and the overlying air supplied buoyancy. The original cast iron cap and connecting wire to the trigger spring are missing. The spring-loaded firing device is in the "fired" position.

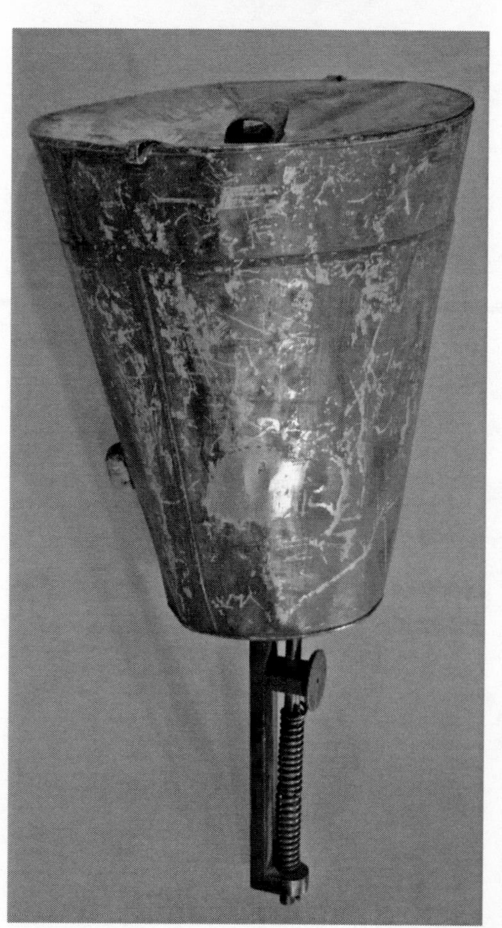

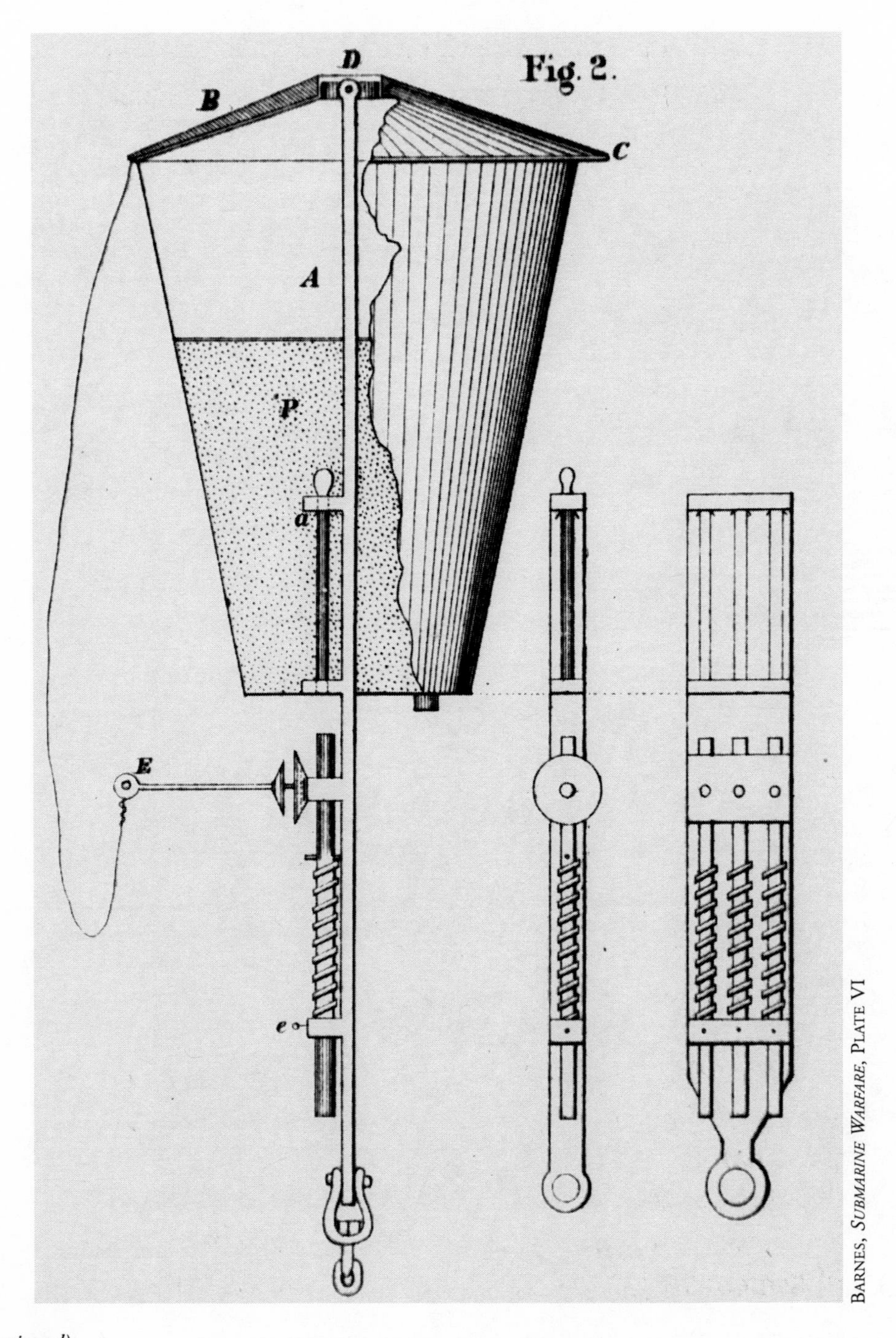

Barnes, *Submarine Warfare*, Plate VI

(continued)

This torpedo was anchored in the river bottom of a bay with a mushroom anchor (see specimen on page 190). When the metal cap (missing) on top of the chamber was knocked off by a passing ship, it would fall and pull the wire connected to the spring-loaded release mechanism, which would activate the trigger powered by a the coiled spring. A hammer would then strike a percussion cap in the bottom of the case, exploding the charge.

These torpedoes were used to sink the Baron de Kalb *on the Yazoo River, Mississippi, and monitor* Tecumseh *in Mobile Bay, Alabama. Many were also found in the James River, Virginia.*

An additional example of a Fretwell-Singer mine is on display in the exhibit adjacent to the gift shop at Patriots Point Naval and Maritime Museum, Mt. Pleasant, Charleston, South Carolina (Images: Herbert M. Schiller).

Drifting Horological Torpedo. *United States Military Academy Museum Accession: 3675*

This specimen was submitted by Lieutenant William W. Burns after the war when he was Supervising Commissary of Georgia and Florida. This was believed to be one of the torpedoes used to attack Federal vessels on the St. John's River, Florida. The timer mechanism has been removed from an eight-day clock. A lever is connected to the clock mechanism so that when a certain time has elapsed, the lever is released, allowing a spring-loaded hammer to fall on a percussion cap mounted on a musket nipple, or cone, discharging into the powder container and igniting the device.

Height: 13.0 inches; Width: 18.0 inches (Images: Herbert M. Schiller).

Rains Wood Floating Anchored Torpedo, opposite page: *United States Military Academy Museum Accession: 3676*

Lieutenant Commander George Bacon submitted this specimen, which was captured in Light House Inlet, Charleston, South Carolina, in August, 1863.

This specimen is a wooden cask, 20 inches long and 13½ inches in diameter. This is the type of torpedo designed by Brigadier General Gabriel J. Rains. Wooden cones, 12½ inches long, have been fastened to each end; there is a fastening loop at each end of the wooden cones. There are metal fittings for percussion fuses at the top and bottom sides of the cask; the actual fuses are missing.

The inside of the keg is coated with pitch to ensure that it is waterproof. Frequently, the exterior of the keg was coated as well. The keg would be partially filled with gunpowder. Enough air would be left to give the keg light buoyancy. By means of the fastening loops, the keg would be anchored to the bottom of the river or bay with rope and one or two mushroom anchors *(see specimen page 190)**.*

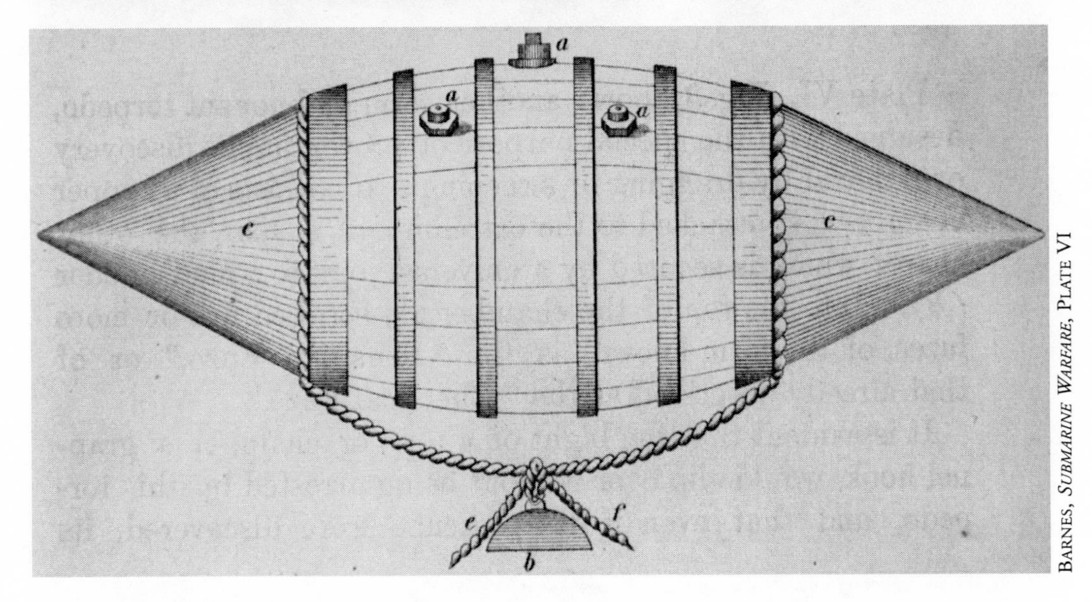

Barnes, *Submarine Warfare*, Plate VI

(continued)

This type of torpedo was easily manufactured by the thousands and was the most popular type with Confederate forces. They were easy to place and very reliable. They frequently had multiple fuses on the top. They were often connected together to form a barrier. A ship striking the percussion primer exploded the charge.

Keg torpedoes, with a single fuse hole and fuse, were used as subterra shells in the defense of Fort Wagner, South Carolina. They were buried with a plank over the fuse to enhance their effectiveness.

The length of this specimen in 20 inches; it is 13½ inches in diameter (Images: Herbert M. Schiller).

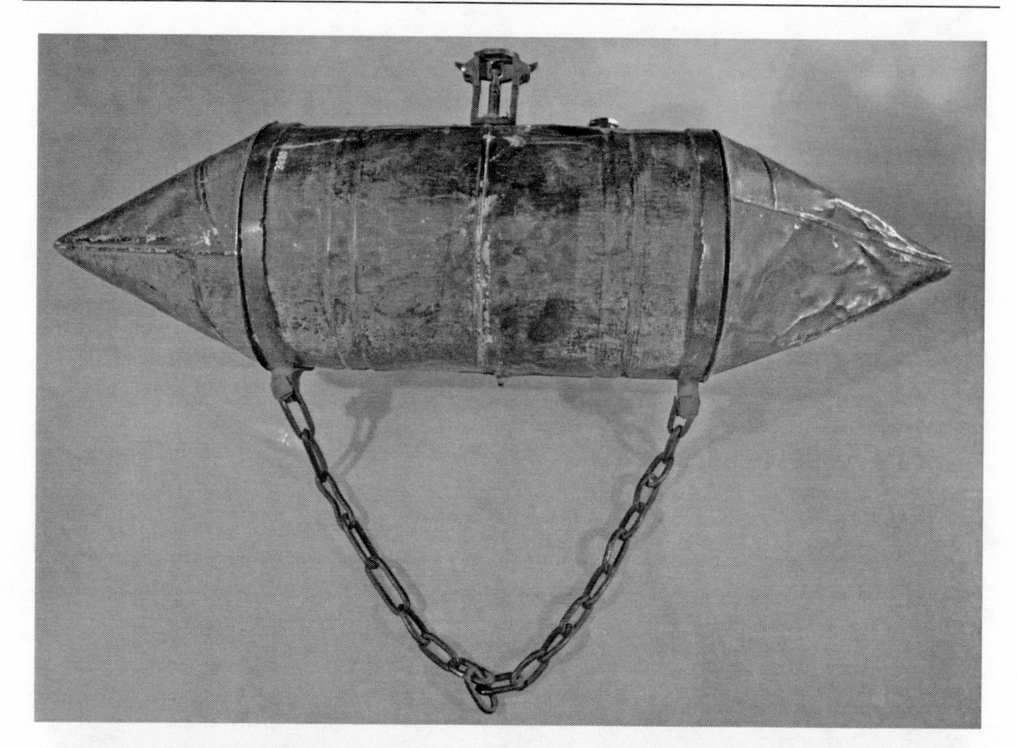

Rains Style Metal Floating Anchored Torpedo. *United States Military Academy Museum Accession: 3680.*

This specimen was donated to the museum by Captain Peter S. Michie. It was captured at the Confederate States Torpedo Factory and presented to the museum in August 1865.

This torpedo is a boom torpedo and consists of a tin cylinder, 19½ inches long by 10½ inches in diameter. The conical projections at either end give it an overall length of 43½ inches. The filler plug is adjacent to the brass fitting for the fuse mechanism.

It was designed to be fired by mechanical percussion, but could be fired by any type of fuse. The wire on the trigger could be connected to a float, to the shore, or to another similar torpedo. A jerk on the wire would release the hammer, which actuated by a coiled spring, would strike the primer, igniting the charge. This was a similar firing mechanism to the Fretwell-Singer torpedo (Images: Herbert M. Schiller).

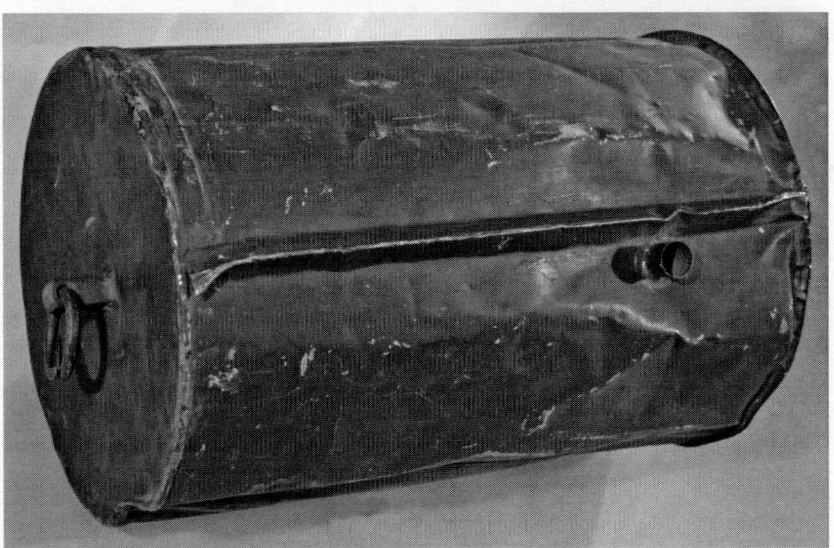

Metal Case Floating Anchored Torpedo, above and following page: *United States Military Academy Museum Accession: 3678*

This specimen was submitted to the museum by Captain Peter S. Michie, who found it in the Confederate Torpedo Factory in Richmond, Virginia, in April, 1865.

It consists of a cylindrical tin case 17 inches long and 11 inches in diameter, which contained gunpowder and air for buoyancy. A metal rod passes through the long central axis and ring bolts are present at either end. A filling port is on the side of the case.

These torpedoes lie on their side in the water, and are held in place by lines through the ring bolts to anchors. Wire passed through a copper disc (missing) and connected within the torpedo to friction primers, similar to those used to discharge artillery pieces. The wire passed out of the container to shore, where it could be pulled by an operator when the vessel passed over the torpedo.

Other configurations had the cylindrical cases connected to one another, so that movement of the line of containers could cause discharge. Another configuration had pieces of floating wood connected by wire to the side of the container, so that if they became fouled in propellers or side-wheels, they would discharge the torpedo.

The notes accompanying this specimen indicate that it was supposedly invented by Zere (or Zedekiah) McDaniel in late 1862 at Vicksburg. It was used frequently in western waters where the connecting trigger wires were concealed by the muddy water (Images: Herbert M. Schiller).

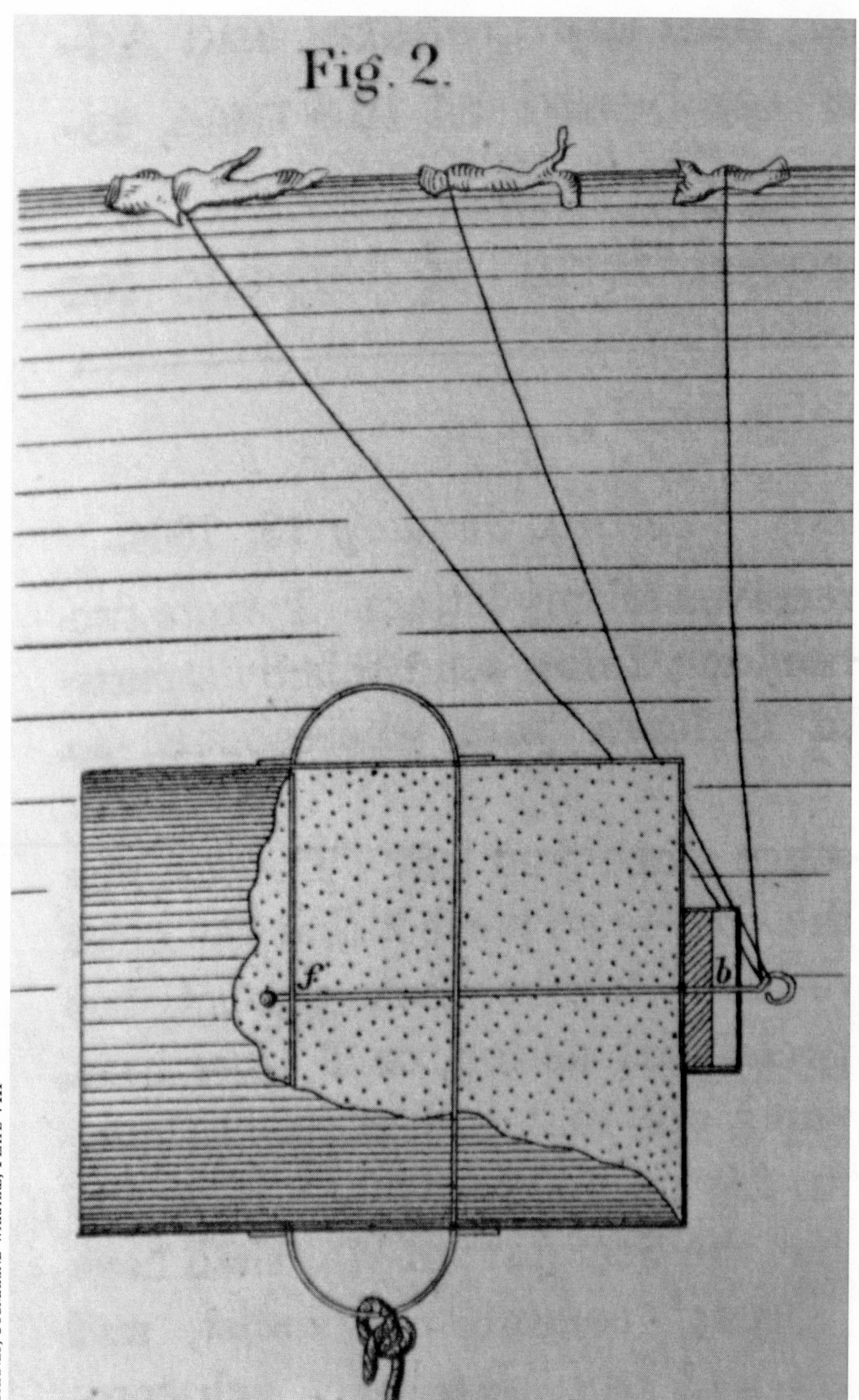

BARNES, *SUBMARINE WARFARE*, PLATE VIII

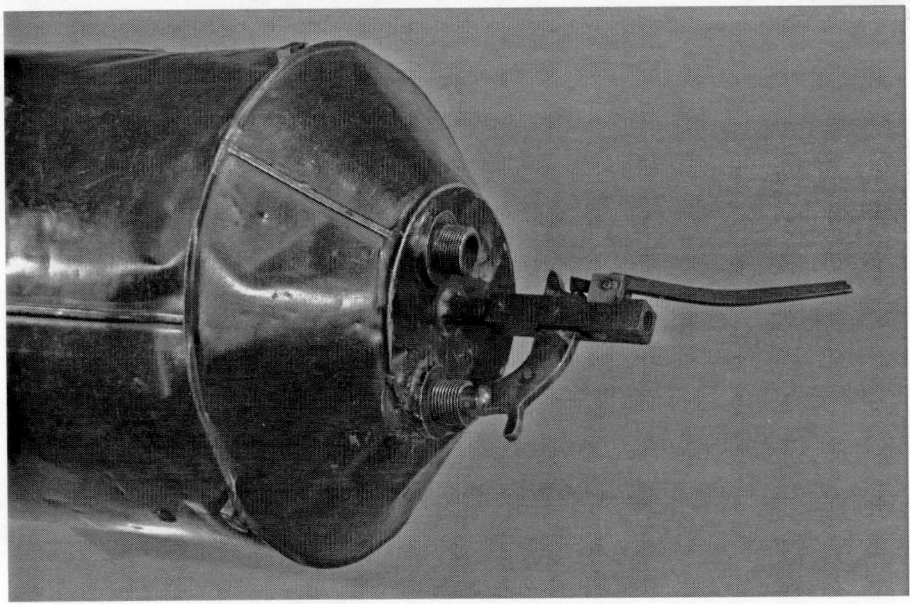

Drift Percussion Propeller Torpedo, above and following page: *United States Military Academy Museum Accession: 3679*

This specimen was submitted to the museum by Captain Peter S. Michie. He found it at the Richmond Torpedo Factory and presented it to the museum in August 1865.

The device is a cylindrical tin case 17 inches long and 12 inches in diameter, with one beveled end. An axial iron rod holding a large brass percussion hammer pierces the axis of the case. Two openings are present at the beveled end. The first contains threads for the sealed percussion device (which is missing). The other opening is the filling port (the cap is missing).

The torpedo was connected to a floating spar or log by means of iron bolts passing through holes in the axial rod outside the cylindrical case. The hammer was actuated by a leaf spring (missing) and long lever pivoted on the rod. The leaf spring was connected to a propeller (missing). The current would carry the torpedo against a ship. When the torpedo was immobilized, the propeller would revolve, which released the lever which, in turn, released the hammer. The hammer, actuated by a spring, would strike the percussion primer and explode the device (Images: Herbert M. Schiller).

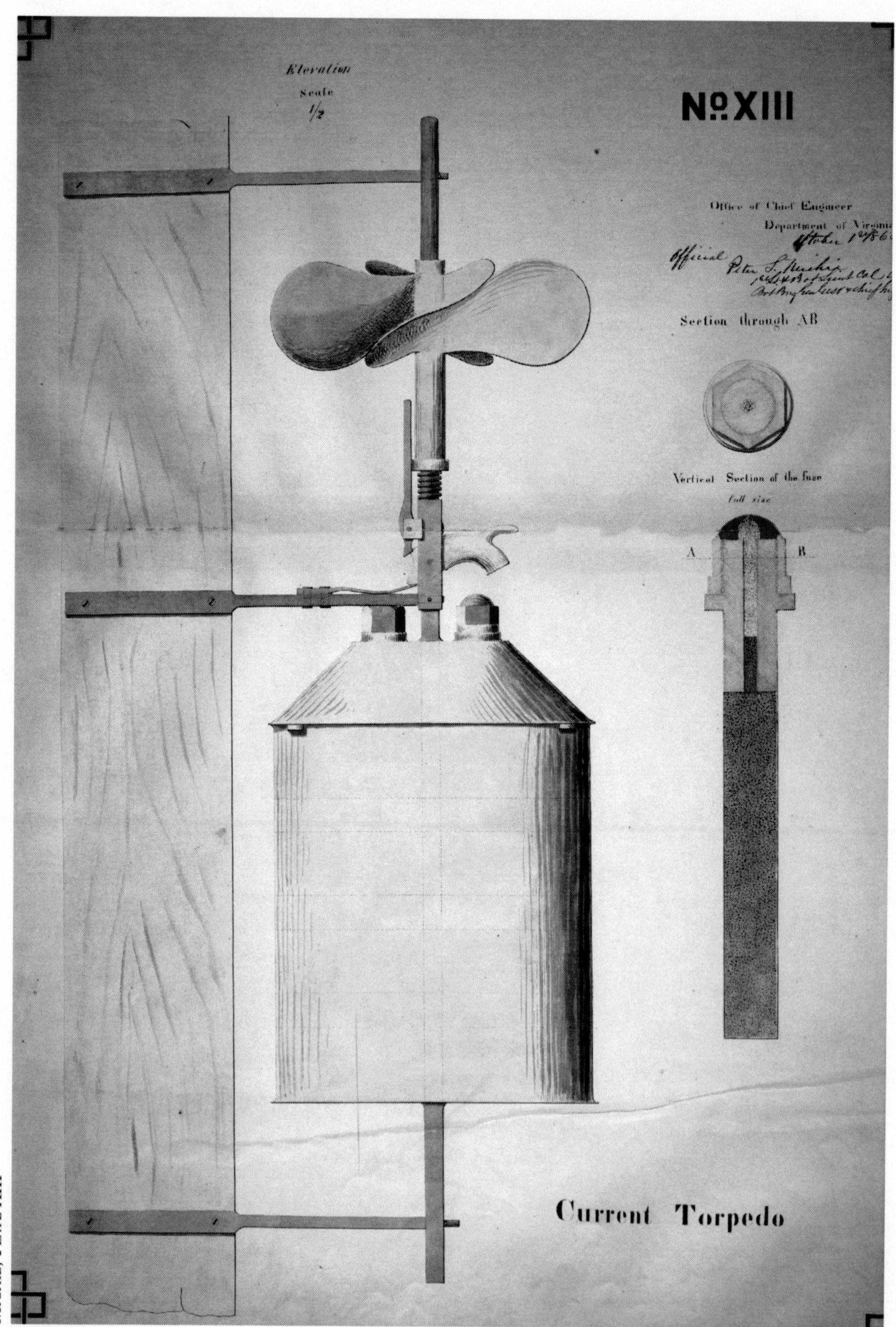

Michie, Plate XIII

Spar Torpedo. *United States Military Academy Museum Accessions: 3681*

This specimen was donated to the museum by Captain Peter S. Michie.

This torpedo was made of sheet copper in an egg-shaped configuration. It is 26 inches long and 13 inches in diameter. The joints are brazed and there are sites for fitting four fuses. The filling port is in the mid-portion. The holes for four fuses are at the end. These torpedoes were attached to a long spar by a basket arrangement (missing).

The spar was attached to the bow of the boat, which would then ram it into the target vessel below the water line. Sulfuric acid fuses and Rains' sensitive fuse primers were used on these torpedoes.

Soda water tanks were widely used to make these torpedoes.

This same design is similar to the one deployed by Lieutenant William B. Cushing to sink the Confederate ironclad Albermarle (Images: Herbert M. Schiller).

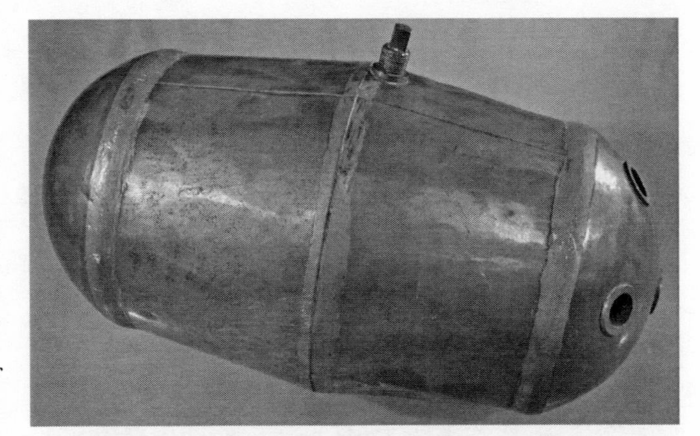

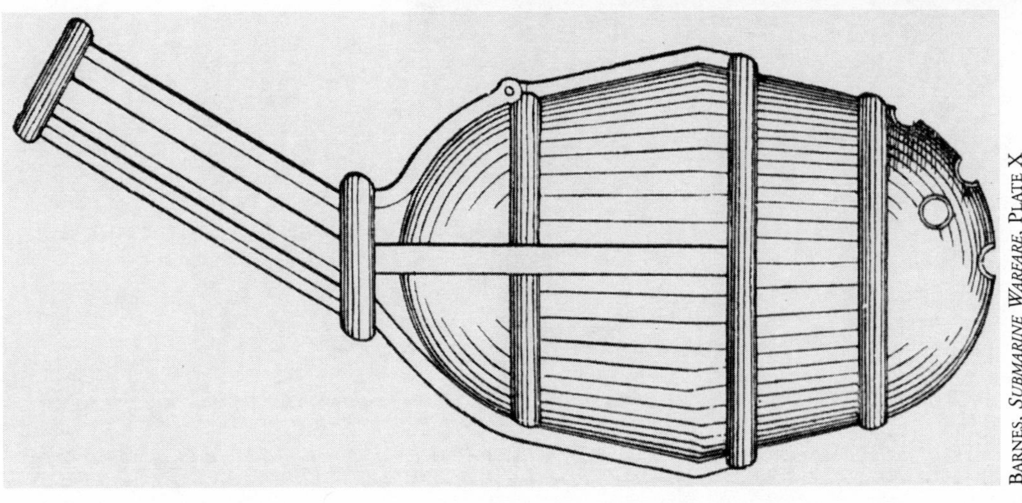

BARNES, *SUBMARINE WARFARE*, PLATE X

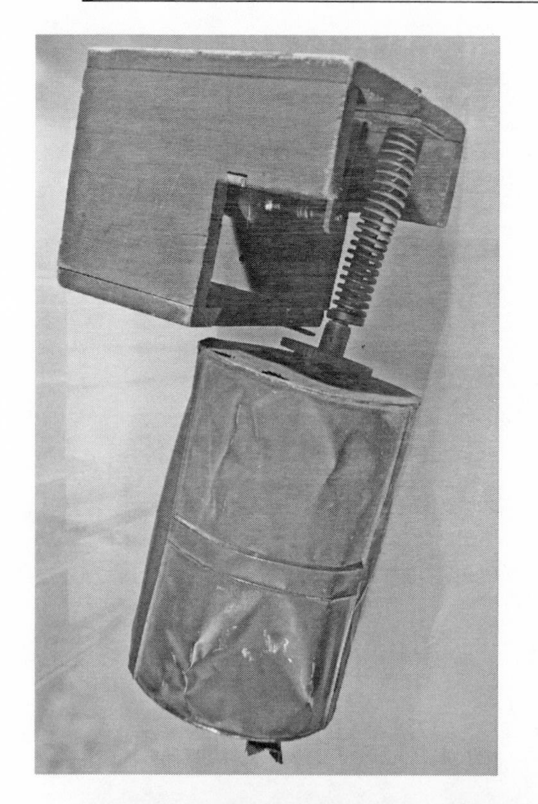

United States Military Academy

Drift Percussion Propeller Torpedo, above and opposite: *United States Military Academy Museum Accession: 3682*

This specimen was donated to the museum by Captain Peter S. Michie. It was captured in the Richmond Confederate States Torpedo Factory in April, 1865, and donated to the museum the following August.

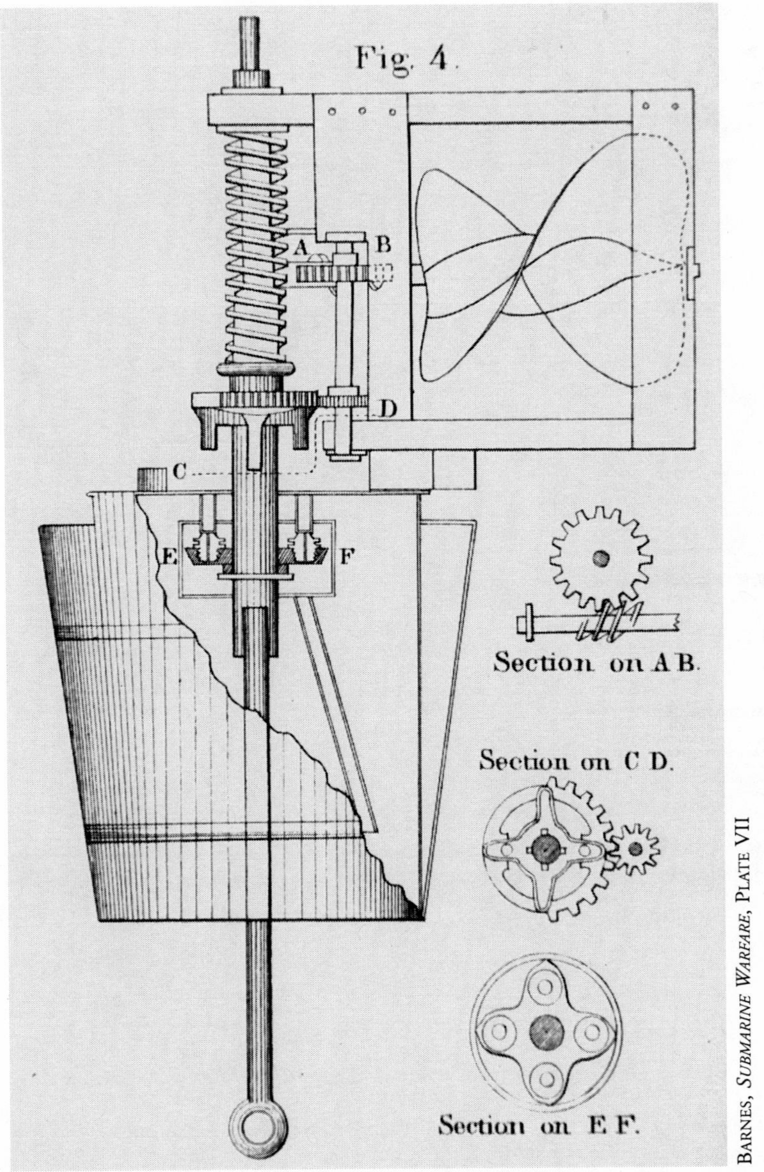

BARNES, *SUBMARINE WARFARE*, PLATE VII

(CONTINUED)

This torpedo consists of a cylindrical tin case 16½ inches long by 11½ inches in diameter and an iron rod through the axis. The upper end of the rod contains a hammer and coil spring. Connected to the rod and hammer by gears is a tin propeller encased in a wooden box, 16 inches by 12 inches by 11 inches. Inside the tank beneath the hammers are two percussion primers supported on an axial rod.

Most of the mechanical machinery portion of this torpedo specimen is missing.

When the torpedo was carried against a vessel and arrested, the tide or current would revolve the propeller, turning the gears, and releasing the hammers. Actuated by a coiled spring, the hammers would strike the percussion primers, exploding the device. The hammer and spring mechanism was similar to the Fretwell-Singer firing system.

A buoy suspended from the ring on the end of the axial rod would keep the torpedo at the proper depth.

These torpedoes were difficult to control and could be actuated by the force of the current alone or when they floated against other obstructions.

Two blades of the propeller are partially visible in the second photograph. The lower photograph shows the box protecting the propeller turned 180° from its operational axis (Images: Herbert M Schiller).

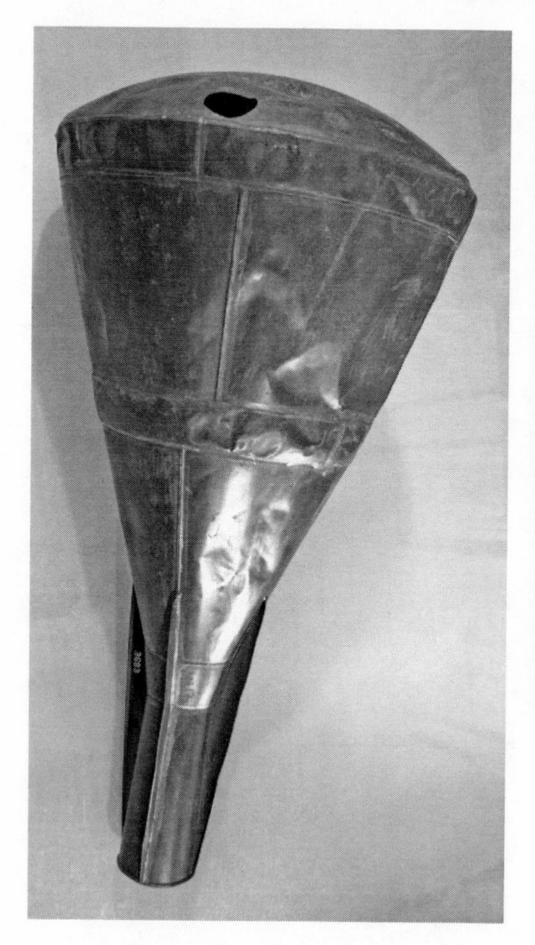

Brooke Swaying Spar Anchored Torpedo, above and opposite: *United States Military Academy Museum Accession: 3683*

This specimen was presented to the museum by Captain Peter S. Michie. It was captured in the Richmond Confederate States Torpedo Factory in April, 1865, and presented to the museum the following August.

This is a swaying (Brooke) torpedo. The conical tin case is 17¼ inches in diameter and is 31½ inches in length. It contains a large chamber for the gunpowder charge and air. The internal air pocket kept the torpedo upright. There are holes for five percussion primers in the dome apex. At the opposite end is a cylindrical collar to attach to a spar.

The spar had a universal joint at its base, and it connected to an anchor. The torpedo was placed below the water line, free to sway with the current. When a vessel bottom strikes the primer, the device explodes.

These torpedoes were one of the most effective Confederate designs. They were planted in rivers and harbors, and were invisible from the surface. Their conical shape and the location of the primers made them difficult to sweep or remove. Often, a line was attached from the swaying torpedo to a firing mechanism (a simple friction primer) of a "turtle" torpedo resting on the bottom in the direction from which a Federal vessel would most likely approach when dragging for torpedoes. If the swaying torpedo was pulled out, it would pull the wire, and friction primer, and the turtle torpedo would explode.

An additional example of a Brooke swaying mine is on display in the exhibit adjacent to the gift shop at Patriots Point Naval and Maritime Museum, Mt. Pleasant, Charleston, South Carolina (Images: Herbert M. Schiller).

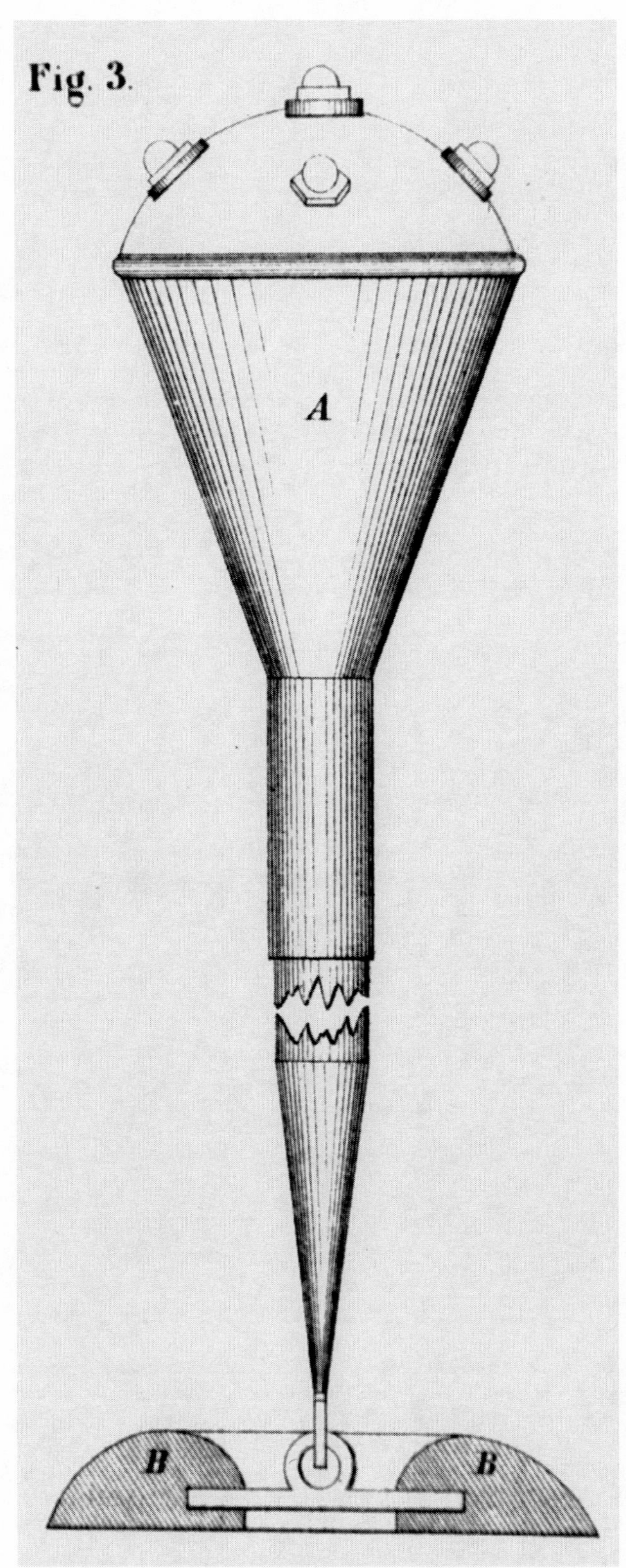

BARNES, *SUBMARINE WARFARE*, PLATE VI

United States Military Academy

Frame Torpedo, above, opposite and page 186. *Top left: Confederate Frame Torpedo removed from Mobile Bay on November 10, 1864. This is the same as Rains' Submarine Mortar Battery (manuscript p. 7) and Michie's Shell Torpedo (manuscript p. 60)(courtesy Naval Undersea Museum, Keyport, WA). Top right: Frame Torpedo specimen in 1950s photograph when it was located at Trophy Point. Bottom: Pair of frame torpedoes, of different manufacture, taken from the Cape Fear River. The one on the right has an external groove to facilitate separation when exploded; the one on the left has an area of thinning cast into the interior. The one on the right also has a filling port. The eye-bolts were probably added by the previous owner, who used them to help anchor a pier. The style of the flanges is also different (courtesy Fort Fisher).*

United States Military Academy Museum Accession: 6515

This type of torpedo consists of a cylindrical iron shell of varying length (22 inches to 28½ inches) and varying diameters (11½ inches to 12¼ inches) with four flanges at the base. The flanges are perforated for bolts to attach the torpedo to a wooden frame.

These torpedoes weigh approximately 400 pounds empty and contained approximately 27 pounds of gunpowder. Although usually filled through the fuse opening, one specimen at Fort Fisher, North Carolina, has a separate filling port.

As noted in Rains' Torpedo Book, manuscript page 8, these shells were cast with a circumferential thinness which allowed the upper portion to project into the contacting vessel. The above photograph shows the shell on page 184, top right, years later after rust has caused separation; the area of thinness is apparent.

The specimen on the right of the bottom image, page 184, shows one of the Fort Fisher specimens to have an external scored groove.

A percussion fuse would be screwed into the tip. The image on page 186 illustrates the abatis-like frame.

Although they could be seen in low water, these were very effective torpedoes. Frame torpedoes were generally placed in shallow or narrow waters with slight tide differences. They were also place on top of sunken hulls in blocked channels. The explosion was caused by a vessel bottom striking the torpedo.

The frame torpedoes were used extensively in the waterways around Charleston, South Carolina. Frame torpedoes damaged the Montauk in the Ogeechee River, Georgia, and severely damaged the Jonquil in the Ashley River, South Carolina.

An additional intact frame torpedo is located in The Charleston Museum, Charleston, S.C. A fragmented specimen is located in Old Fort Jackson, Savannah, Georgia (Image: Herbert M. Schiller).

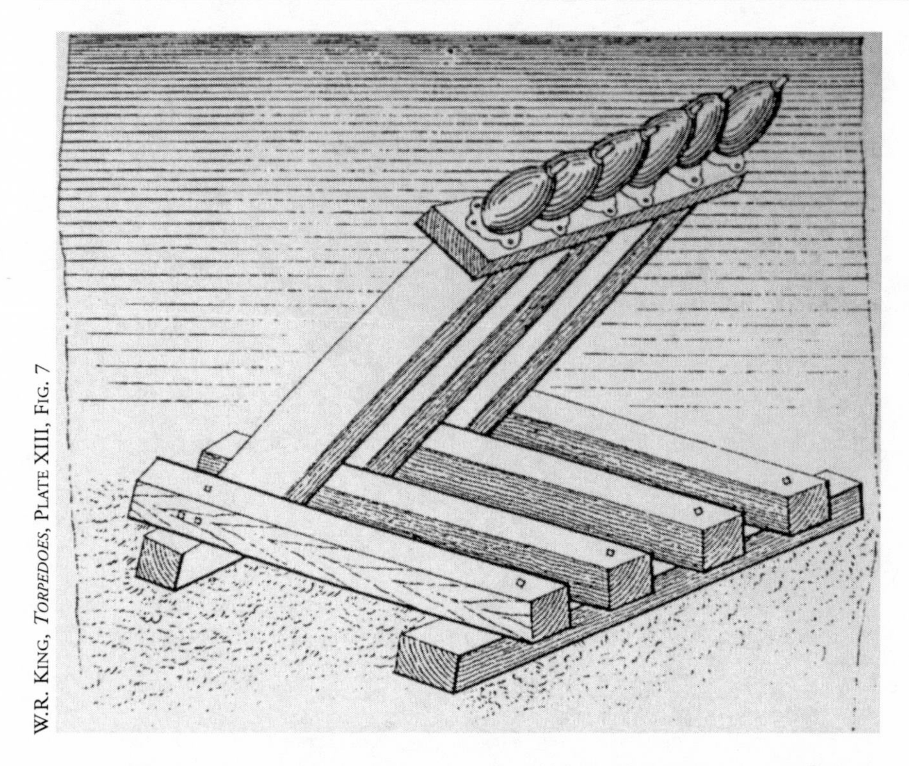

W.R. King, *Torpedoes*, Plate XIII, Fig. 7

Large Boiler Galvanic Torpedo. *United States Military Academy Museum Accession: 6517*

The donor of this specimen in unclear, but may have possibly been Captain Peter S. Michie, who sent an "electric submarine tank" in August, 1865.

This torpedo is made of ¾ inch boiler iron; the tank is 78 inches long and 45 inches in diameter. Two iron lifting lugs are riveted to the body.

These tanks were fastened to parallel beams and lowered into the river. The torpedo contained approximately 2,000 pounds of gunpowder. The fuse diagrams are shown in both the Rains and Michie plates. Two copper wires connect with a goose quill filled with fulminate of mercury at the center of the case. Fine platinum wire runs through the quill connecting the two copper conducting wires. The quill is surrounded by gunpowder. The electric charge from the battery heats the platinum wire, which detonates the fulminate (which detonates at a relatively low temperature) and explodes the torpedo.

Although boilers were initially used for these large galvanic torpedoes, they were later specially fabricated.

These torpedoes were developed by the Confederate Navy at the Submarine Bureau. Such a torpedo was responsible for the destruction of the Commodore Jones *in the James River on May 6, 1864. Galvanic torpedoes were used in Charleston, South Carolina, and on the James River, Virginia, the Cape Fear River, North Carolina, as well as elsewhere (Image: Herbert M. Schiller).*

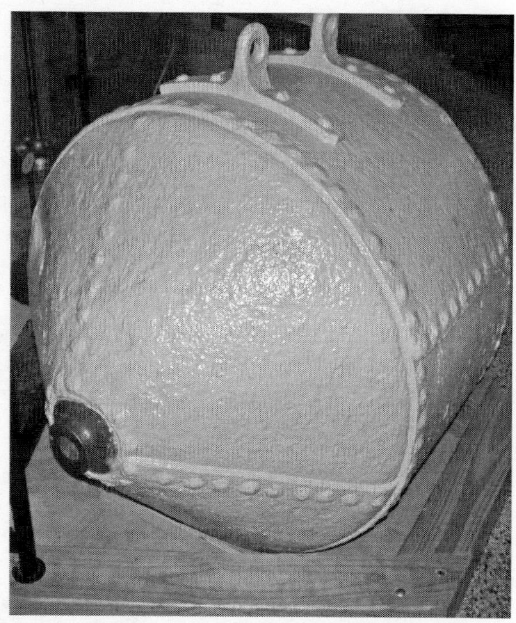

Medium Boiler Galvanic Torpedo. *LEFT: United States Military Academy Museum Accession: 6516.*

The galvanic torpedo is made of ¾ inch riveted boiler iron formed into a cylindrical tank with conical ends. It is 42½ inches long and 30 inches in diameter. Lifting lugs are riveted to the tank. The loading port is visible on one end; the plug is missing. The wire and fuse mechanism entered through the hole at the opposite conical end. The USMA records estimate this torpedo would hold 100 to 150 pounds of gunpowder; I suspect it would hold considerably more (Image: Herbert M. Schiller).

RIGHT: This galvanic torpedo was discovered in the beach sand immediately north of the north wall of Fort Fisher, North Carolina, in 1964 (see Stanley South, An Archaeological Education *[New York: Springer Science—Business Media, 2005] pp, 175–177). It is 49 inches long and 36 inches in diameter (courtesy Fort Fisher).*

LEFT: ***Trunion Galvanic Torpedo.*** *United States Military Academy Museum Accession: 3669*

The history of this specimen is unknown. It is 20 inches wide (without measuring the trunnions) and 29 inches high. The trunnions on the sides were probably used for raising and lowering the torpedo. There is a screw plug for the fuse wires at one end; the iron bolt, presumably the loading port, is at the opposite end.

The records suggest that this torpedo may be been used for clearing obstructions by lowering it from a ship and exploding it (Image: Herbert M. Schiller).

Rains Sensitive Fuse and Safety Cap. *Top: This Rains sensitive fuse with safety cap has a large hexagonal nut for tightening it into the body of the torpedo (Image: Atlanta History Center).*

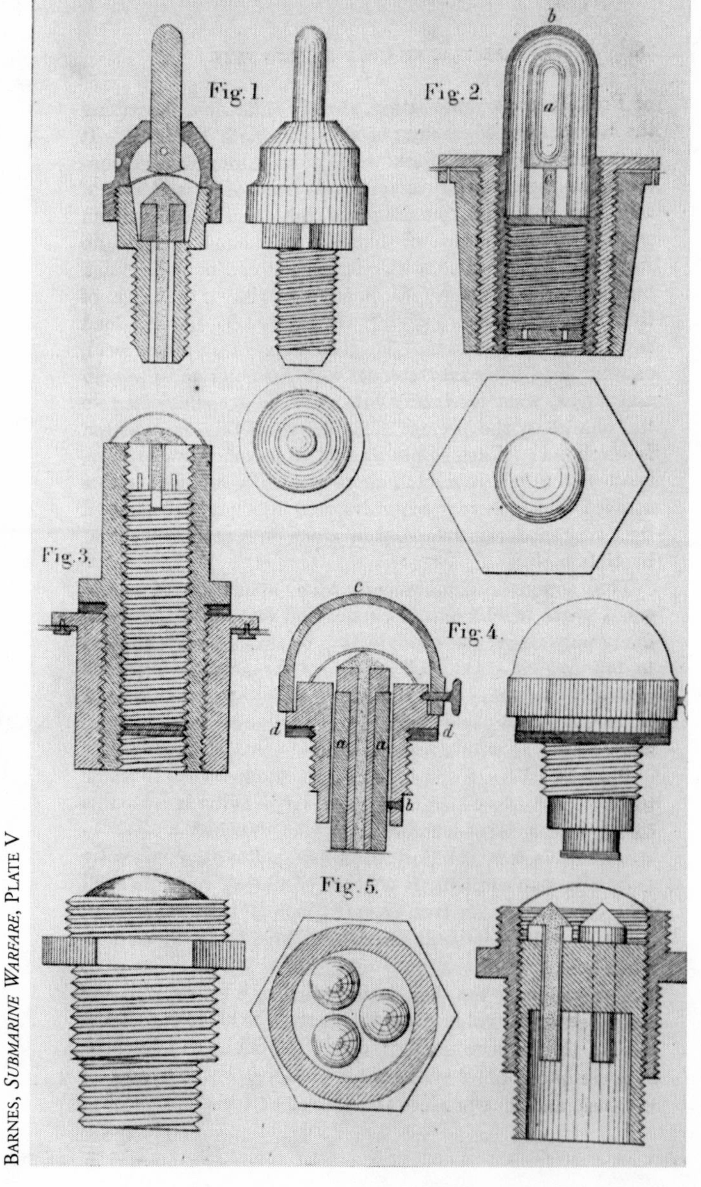

Barnes, *Submarine Warfare*, Plate V

Contact Fuses: Figure 2: Confederate Captain Francis D. Lee's Chemical (sulfuric acid) Torpedo Fuse; Figure 3, Captain Lee's Sensitive Torpedo fuse; Figures 1, 4, 5: Variants of Rains' fuses.

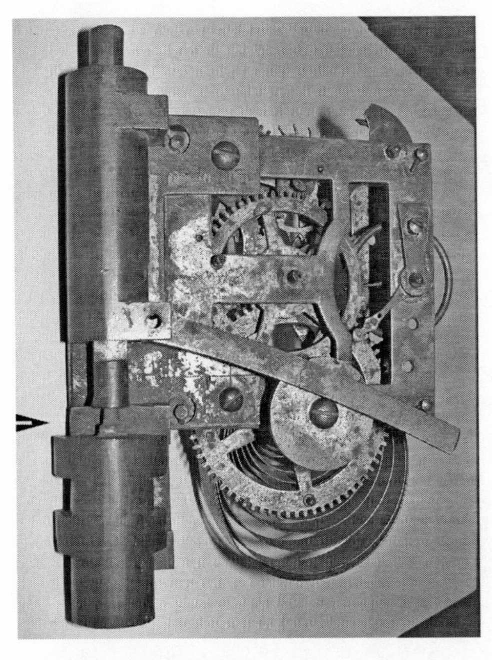

Horological Torpedo Mechanism. *Mechanism for a horological torpedo like the one used by John Maxwell at City Point, Virginia.*

The 8-day ordinary clock mechanism was connected to a spring-loaded rod in the upper cylinder, which, when released, would detonate a percussion cap (not present) on the musket nipple, or cone (to the right of arrow) of the lower cylinder, exploding the associated charge. The rod is in the "down" position on top of the nipple. The mechanism is $5\frac{5}{8}$ *by* $4\frac{3}{16}$ *by* $3\frac{3}{16}$ *inches.*

Compare with Michie's Plate XVII, which must have been based on an available specimen at the Torpedo Bureau laboratories in 1865.

The specimen was presented the Petersburg National Battlefield by Alice Evans Hackett, the granddaughter of John Maxwell, in 1990 (Image: Petersburg National Battlefield).

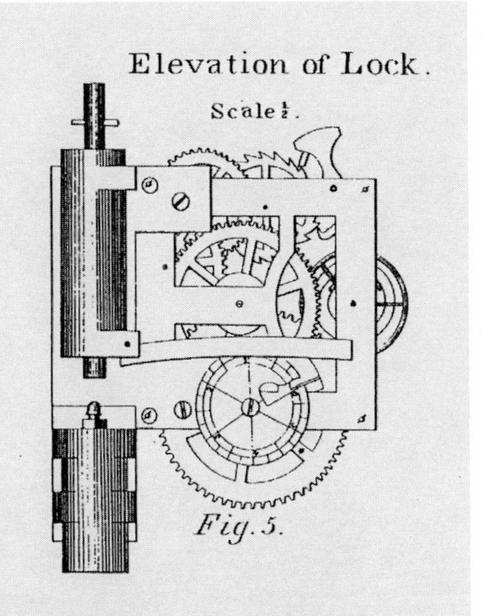

Section through axis of hammer and nipple.

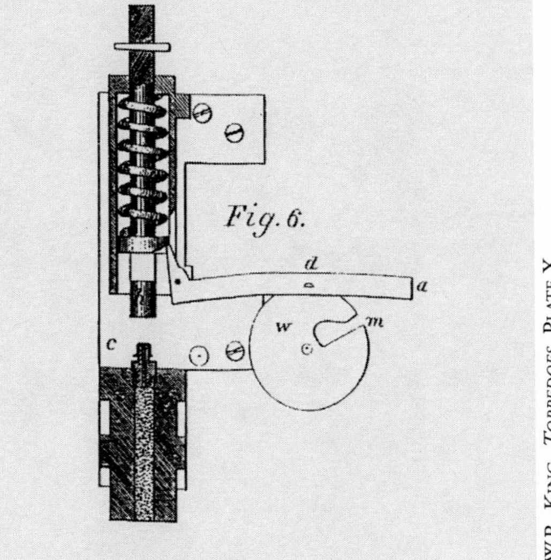

W.R. KING, *TORPEDOES*, PLATE X

The clockworks turn the wheel (w) fixed on the axis of the mainspring. The rotation of this wheel brings the recess (m) under the pin (d) which drops into it, letting the lever arm (a) descend and releases the hamper (h) which is impelled by the spring and strikes and explodes the cap (c) and explodes the torpedo.

Mushroom Anchor. *Mushroom anchors were used to anchor various submerged torpedoes. This specimen was taken from the Cape Fear River, North Carolina. It is 20 inches in diameter and 5 inches high. The ring bolt is 2½ inches in diameter (courtesy Fort Fisher).*

Photograph, possibly by George Barnard, taken in April or May, 1865, at the Charleston Arsenal, on the west side of Ashley Avenue approximately midway between Doughty and Bee Streets, facing northwest.

A: *Wood Floating Anchored Torpedoes,* ***B:*** *Confederate Submarine Mortar Shells (Frame Torpedo Shells),* ***C:*** *Confederate Arm Anchored Torpedo (the detonating device is the spoke-like arrangement on the top [Michie manuscript page 70]) (Image: Library of Congress DIG-CWPB-02383).*

Confederate Torpedoes, taken from the James River. This image, taken late in 1864, was on the James River near the site of the incomplete Dutch Gap Canal excavation. The two seated men inspect a Confederate keg torpedo. Under the shed are metal case floating anchored torpedo containers awaiting filling for use as mines in the river. (Image: United States Army Military History Institute, Mollus Collection, volume 27, p. 1338).

Editor's Appendix 3: Plates Not Referenced in Rains' Text

[Plate, 21½]

For a non metallic primer to be used in firing cannon between decks.

Cut off pieces of quick match 6 inches long, and wrap the single strand around the wood of a common friction match put then together in water a moment (the match-end slightly projecting) (and roll on paper with paper) & wrap paper around the three, the match, the moistened cotton strand, & gunpowder sticking and paste its edge down.

The match now must have a good size pin run through it & cut off, so as to let it enter the vent of a cannon as far as that at right angles. The quick match when making, had better be dusted with fine powder, and it must be brought up nearly to the top of the match in the primer. To fire this primer—insert it to the pin, and with an elastic piece of wood as a lath with a piece of sand paper fastened on the end, press that on the match and pull toward you—the friction fires. These matches were first used at Baton Rouge, La., to fire a salute to Genl. Zac. Taylor by Col. Boneville, to whom I gave them. They were very successful.

The Hirsh Screw Propeller adopted by the British Navy

[Page of text following plate 23½]

A new explosive

Dissolve compressed gun cotton in nitro-glycerine which becomes gelatinous or gummy not explosive. Add 4 per cent camphor to 96 of gelatin and which consists of 90 per cent of nitro-glycerine & 10 of [illegible]-cotton. The gelatinous mass is elastic, transparent of pale yellow color which can be cut with a knife but cannot be fired by a blow. When cotton fibre is subjected to the action of surphuric acid a white pulverulent substance is obtained named hydrocellulose—is easily soluble in nitric acid when it becomes nitro-hydrocellulose. This compound mixed in the proportion of 40 per cent with 60 per cent nitroglycerine forms the most powerful means of ignition ever hitherto discovered. By properly constructed firing cartridges of that substance the explosive gelatin becomes as manageable as ordinary powder with less danger and far more force.

[faint pencil sketch of compressed air torpedo]

The compressed air torpedo: The cigar shaped fish torpedo gets its motion from compressed air stored inside and this, if issuing at the tail, sets in motion a screw which revolves with considerable velocity. A well constructed fish torpedo will run many hundred yards at a velocity of twenty miles an hour, and on striking its head which contains the charge, explodes with considerable violence. The fish torpedo is therefore a weapon of terrible effect.

My own propelling force:

Compressed air
Frozen carbo. acid
Liquefied ammonia
Sulphuric acid & carb. lime

Nitrolin

From 5 to 20 parts sugar or syrup are mixed with from 25 to 30 parts nitric acid in a wooden or gutta percha vessel. Of this compound 25 to 30 parts are mixed with 13 to 35 parts of nitre & from 13 to 15 parts of cellulose.

Notes

I. Gabriel J. Rains

Biographical Introduction

1. An excellent biography of Gabriel Rains is William Davis Waters' 1971 Wake Forest University master thesis *Gabriel J. Rains: Torpedo General of the Confederacy* (hereinafter, Waters, *Thesis*). Waters later modified it in "'Deception Is the Art of War': Gabriel J. Rains, Torpedo Specialist of the Confederacy," *The North Carolina Historical Review*, vol. 66, Number 1, January 1989, pp. 29–60 (hereinafter referred to as *NCHR*). Milton F. Perry's *Infernal Machines* (Baton Rogue: Louisiana State University Press, 1965), offers an account of the use of these devices on land, in rivers, and at sea. Timothy S. Wolters' "Electric Torpedoes in the Confederacy: Reconciling Conflicting Histories," *Journal of Military History*, vol. 72, Number 3, pp. 755–783, offers an account of the development of the Confederate Army and Navy programs; Gertrude Carraway, "North Caroliniana," Charlotte *Observer*, January 11, 1942.

2. Gabriel J. Rains Papers, July 30, 1840, South Carolina Historical Society.

3. Waters, *NCHR*, pp. 31–32; Letter June 21, 1860, Gabriel J. Rains Papers, United States Military Academy.

4. *The War of the Rebellion: A Compilation of the Official Records of the Union and Confederate Armies* (Washington: Government Printing Office, 70 volumes, 1880–1901) Series 1, vol. 51, part 2, p. 306 (hereinafter referred to as *ORA*).

5. *ORA*, Series 1, vol. 11, part 3, p. 136.

The development of torpedoes has a history going back to at least the 18th century. Rains' work, as well as that of Matthew F. Maury and Hunter Davidson, did not occur *de novo*, but rather built on work of their predecessors.

Richard Delafield's *Report on the Art of War in Europe in 1854, 1855, and 1856* (Washington: George W. Bowman, 1860), pp. 109–122, and Plates 13 and 14 details the use of contact fuses in subterra shells and the use of electric discharge for single or multiple mines on land or underwater during the Crimean War by the Russian military. He also reviews advances made by other European countries in the field of torpedo warfare.

John S. Barnes, in his *Submarine Warfare, Offensive and Defensive, Including a Discussion of the Offensive Torpedo System, Its Effects upon Iron-Clad Ship Systems, and Influence Upon Future Naval Wars* (New York: Van Nostrand, 1869), pp. 17–60, recounts the use of torpedoes in America from the Revolutionary War, the War of 1812, Samuel Colt's experiments in the early 1840's, and the use of both land an war torpedoes in the Crimean War, the last abstracted from Delafield's report.

6. *ORA*, Series 1, vol. 11, part 3, pp. 516–517. Rains, in a July 5, 1862, letter to his brother George Washington Rains mentions, in the context of the frame torpedo shell he was developing, that he was using his "sensitive primer tubes" to detonate them. (Gabriel J. Rains Papers, July 5, 1863, South Carolina Historical Society.)

I assume Gabriel Rains had already developed the sensitive fuse prior to using them in May 1862. I assume what he used in the Seminole War was some modification of the friction primer used to fire artillery pieces.

7. *ORA*, Series 1, vol. 11, part 3, pp. 509–510.

8. *ORA*, Series 1, vol. 11, part 3, p. 510.

Rains notified Randolph that he preferred to be appointed to the river defenses. On July 5, 1862, he wrote his brother George that he was anxious to get from under the command of Generals Longstreet and Hill, "especially Longstreet who is a drinking man & said to be an inebriate." In addition, Rains had ranked Longstreet in the "old army." (Gabriel J. Rains Papers, July 5, 1862, South Carolina Historical Society.)

9. *ORA*, Series 1, vol. 11, part 1, pp. 943–945.

10. *ORA*, Series 1, vol. 11, part 3, p. 608; G.J. Rains, "Torpedoes," *Southern Historical Society Papers*, vol. 3, 1877, p. 260.

11. Perry, *Infernal Machines*, pp. 3–4.

12. *ORA*, Series 1, vol. 7, p. 534.

13. *Official Records of the Union and Confederate Navies in the War of the Rebellion* (Washington: Government Printing Office, 30 volumes, 1894–1914), Series 1, vol. 7, p. 544 (hereinafter referred to as *ORN*), report includes map and diagrams.

In a letter to his brother from Drewry's Bluff, Rains wrote that there were eleven "magazines" containing 150 pounds of gunpowder and two containing 2,000 pounds of gunpowder in the James River, diagramed their location, and stated they were designed to be discharged with a Wollaston battery. Two had washed away in a recent freshet. (Gabriel J. Rains Papers, July 5, 1862, South Carolina Historical Society.)

14. *ORN*, Series 1, vol. 7, p. 546.

15. Gabriel J. Rains "Torpedoes," unpublished notebook, The Museum of the Confederacy, Richmond, Virginia, manuscript pp. 1–12. (hereinafter referred to as Rains, *Torpedo Book*); *ORA*, Series 1, vol. 42, part 3, p. 1219.

16. Chris Fonvielle, personal communication, July 17, 2009.

17. ORN, Series 1, vol. 7, p. 61; Perry, *Infernal Machines*, p. 33.

18. Waters, *NCHR*, p. 41. Waters notes that the term "Torpedo Bureau" did not appear on correspondence until June, 1864. Perry notes that men assigned to the Bureau had to take an oath not to reveal any information about Confederate torpedoes (Perry, *Infernal Machines*, p. 31).

19. Waters, *NCHR*, p. 42.

20. J.B. Jones, *A Rebel War Clerk's Diary* (New York: Old Hickory Bookshop, 1937), p. 245; Schafer, *Confederate Underwater Warfare*, p. 79; Rains, *Torpedo Book*, manuscript pp. 78–84; "The History of Landmines" at http://members.iinet.net.au/~pictim@iinet.net.au/mines/history/history.html.

Review of the Confederate Patent Office records in The Museum of the Confederacy reveals no application for a patent for the sensitive primer fuse. Hunter Davidson later wrote that Rains "was daft on sensitive fuses, and his experiments were generally disastrous." (Hunter Davidson, "The Electrical Submarine Mine, 1861–1865," *Confederate Veteran*, XVI, 1908, p. 459.)

21. Louis S. Schafer, *Confederate Underwater Warfare* (Jefferson,: McFarland & Company, 1996), p. 76; *ORA*, Series 1, part 2, vol. 52, p. 487.

In this letter Davis infers that the manuscript he had received was a copy. Davis' copy has been lost. The manuscript at The Museum of the Confederacy, as will be seen, may include part of the original, especially some of the plates, but internal evidence shows that portions of it were written much latter. Also see Waters, *NCHR*, p. 45, footnote 47.

22. *ORA*, Series 1, vol. 52, part 3, p. 487.

23. *ORN*, Series 1, vol. 16, p. 385.

24. *ORN*, Series 1, vol. 16, p. 386.

25. *ORA*, Series 1, vol. 18, part 1, p. 1082–1083.

26. *ORA*, Series 1, vol. 24, part 1, p. 200.

27. *ORA*, Series 1, vol. 26, part 2, p. 536; *ORA*, Series 1, vol. 24, part 2, p. 536; Rains, *Torpedo Book*, manuscript page 77.

On September 15, 1863, E.O. Singer, J.P. Fretwell (both experienced designers of torpedoes) along with two other men were ordered to report to Lieutenant General E. Kirby Smith, commanding the Trans-Mississippi Department, "on the special service of destroying the enemy's property by torpedoes and similar inventions." (*ORA*, Series 1, vol. 22, pt. 2, p. 1017.)

28. *ORA*, Series 1, vol. 26, part 2, p. 136; *ORA*, Series 1, vol. 24, part 1, p. 200; *ORA*, Series 1, vol. 26, part 2, pp 136, 180.

29. Rains, "Torpedo Book," manuscript pp a–c, k, l, 6, 11–12, 29, 31, 85, 86, 102–104; *ORA*, Series 1, vol. 28, part 2, p. 324.

30. Waters, *NCHR*, p. 50.

31. *ORA*, Series 1, vol. 32, part 2, p. 738.

32. *ORA*, Series 1, vol. 32, part 2, p. 738; *ORA*, Series 1, vol. 39, part 1, pp. 433–434.

33. *ORN*, Series 2, vol. 2, p. 634; *ORA*, Series 1, vol. 36, part 3, p. 883.

34. Waters, *NCHR*, p. 52; Rains, *Torpedo Book*, manuscript p. 39; *ORA*, Series 1, vol. 42, part 1, p. 954–956.

35. Barnes, *Submarine Warfare*, p. 75.

36. Waters, *NCHR*, pp. 47–49; Waters, *Thesis*, p. 73–74, Rains; *Torpedo Book*, manuscript p. 56.

37. George B. Sandford, *Fighting Rebels and Redskins* (Norman: University of Oklahoma Press, 1964), p. 234; S.H. Nowlin, "Capture and Escape of S.H. Nowlin, Private Fifth Virginia Cavalry," *The Southern Bivouac*, vol. 2, Oct. 1883, p. 70; Henry R. Pyne, *History of the First New Jersey Cavalry* (Trenton, 1871), p. 242; Gordon C. Rhea, *To the North Anna River* (Baton Rouge: Louisiana State University Press, 2000), pp. 41, 50–51.

38. *ORA*, Series 1, vol. 42, part 3, p. 1181.

39. *ORA*, Series 1, vol. 42, part 3, pp. 1181–1182.

40. *ORA*, Series 1, vol. 42, part 3, pp. 281–282.

41. *ORA*, Series 1, vol. 42, part 3, pp. 1181–1182.

42. *ORA*, Series 1, vol. 42, part 3, pp. 1219–1220.

43. Waters, *NCHR*, p. 54; Rains, *Torpedo Book*, manuscript p. 56.

44. Waters, *NCHR*, p. 52; *ORA*, Series 1, vol. 42, part 3, pp. 1219–1220.

45. *ORA*, Series 1, vol. 44, pp. 79, 88, 100; William T. Sherman, *Memoirs of General William T. Sherman* (New York, The Library of America, 1990), p. 670.

46. *ORA*, Series 1, vol. 46, part 1, p. 408; William Lamb, "The Defense of Fort Fisher," *Battles and Leaders of the Civil War.* (New York: The Century Company, 1884), vol. 4, pp. 649, 650, diagram p. 645; *ORA*, Series 1, vol. 28, part. 1, pp. 204, 470; part 2, pp. 78, 213; Stephen R. Wise, *Gate of Hell.* (Columbia, University of South Carolina Press, 1994), p. 134.

47. Rains, *Torpedo Book*, manuscript p. 54; *ORA*, Series 1, vol. 38, part 4, p. 579.

Sherman directed that when torpedoes were found in the Union army's rear, they could be "tested by wagonloads of prisoners, or if need, by citizens implicated in their use.... Of course an enemy cannot complain of his own traps."

An electronic search for "torpedoes" and "subterra" on a CD-ROM of the *ORA* and *ORN* (Guild Press of Indiana, Inc., Carmel, IN, 1996, 1999) reveals extensive use of both land and water devices in all theaters throughout most of the war; an in-depth study of the use of marine and land torpedoes is waiting to be written.

48. *ORA*, Series 1, vol. 42, part 3, pp. 281–282.

49. Waters, *NCHR*, p. 57.

50. Several of the torpedoes described in the manuscript, i.e., the demijohn torpedo, were developed by others. An excellent account of the various types of torpedoes, their construction, methods of detonation including an description of the various electric spark generators may be found in Michael P. Kochan and John C. Wideman's *Torpedoes: Another Look at the Infernal Machines in the Civil War* (self published, 2004; Michael Kochan can be reached at 25 Sunset Drive, Paoli, PA 19301).

51. Gabriel J. Rains, "Torpedoes," *Southern Historical Society Papers*, vol. 3, 1877, pp. 255–256. Rains claimed that 58 Union vessels were sunk by Confederate torpedoes (Rains, *SHSP*, vol. 3, p. 256).

52. Patent number 136, 938; see document at http://www.google.com/patents?id=D4NKAAAAEBAJ&pg=PA1&dq=gabriel+rains&source=gbs_selected_pages&cad=0_1#PPA2,M1

53. John Y. Simon, editor, *The Papers of Ulysses S. Grant*, Vol. 25, 1874 (Carbondale: Southern Illinois University Press, 2003), pp. 422–423. The paths of

Rains and Grant must have crossed in the west prior to Grant's resignation from the army in 1854.

54. National Archives, Microfilm Publication M2048, War Department Letters and Telegrams Sent Relating to the United States Military Academy, 1867–1904, Roll 1, Volume 3: July 1, 1874-October 13, 1875, p. 92.

55. Waters, *NCHR*, p. 59.

Torpedo Book

1. "Remarkable History of a Torpedo Boat," *Scientific American*, June 9, 1866, vol. XIV, p. 406.

2. Rains seems to credit one of his torpedoes with sinking the *Housatonic*, not the *Hunley*.

3. This is the principle of the Jacobi fuse, used by the Russians in the Crimean War, and subsequently used in the contact fuse on spar torpedoes.

4. This statement was made by Capt. (so called) Gray falsely & without a knowledge of facts. Gray was afterwards imprisoned for forgery.—G.J.Rains

[M. Martin Gray seems to have been imprisoned in August, 1864, for embezzling by buying rope cheaply and selling it to the Confederate government at a large profit (Perry, *Infernal Machines*, pp. 166–167).

After hostilities ended, Gray offered his services to the Union Navy, stated that he had directed "for some time of the fabrication and placing of torpedoes" but had become "suspected by the rebels" and spent the last six months imprisoned. The nature of his duties was substantiated by copies of his correspondence which had been found in one of the offices. (J.A. Dahlgren, *ORN*, vol. 16, p. 380.)

5. Probably the Confederate flag-or-truce steamer *Schultz*; see Appendix 1, February 17, 1865.

6. see manuscript page 74.

7. See manuscript page 85X.

8. Having been recently assigned to river defense and stationed at Drewry's Bluff, Rains wrote his brother George that "I have been making some mortar shells." In the letter he sketched a shell similar to that in Plate 3, except it does not show the flanges at the base. At the tip they had "one of my sensitive priming tubes covered with a thin copper ... and soldered water tight." The shell itself "is fastened to a float submerged" The shells were four inches thick, but the upper end has an area of thinness one inch thick so that when the shell exploded, the anterior conical end, in contact with the vessel bottom, separates and penetrated the hull. (Gabriel J. Rains Papers, July 5, 1862, South Carolina Historical Society.)

9. Probably the *Marion*; see Appendix 1, April 6, 1863.

10. The *Weehawken* was slightly damaged, not sunk, on April 7, 1863, by a torpedo; see Appendix 1.

11. Rains did not believe the *Hunley* sunk the *Housatonic*.

12. Emile Lamm, "Ammonia as a Motive Power for Street Cars," *Scientific American*, vol. XXV, Nov. 4, 1871, p. 290.

13. By a camera obscura if convenient.—G.J. Rains

14. Actually two vessels, the gunboat *Otsego* and tug *Bazely*; see December 9 and 10, 1864, Appendix 1.

15. This is an excellent plan to frustrate taking them up.—G.J.Rains

16. Ordered on 11th March 1865 after many of these torpedoes were set in Mobile Bay they exploded daily from the tide leaning them. I think Admiral Taylor says only that torpedoes kept the U.S. Navy out of Charleston.—G.J.Rains

17. See August 9, 1864, Appendix 1.

"The wharf boat at Mound City, containing the reserve supplies of ammunition and stores for Admiral Porter's fleet, was also destroyed by a similar contrivance." Barnes, *Submarine Warfare*, p. 75.

18. Arranged fig. 1 on plate 1st with the female screw also of brass soldered on as stated so that the torpedo can be kept ready with its load of powder if necessary & each screw hole plugged with a wooden stopper, until about to be used when that can be screwed out, and the prepared sensitive primer's brass plug screwed in, its helix 'wet' with tar or pitch. Usually, however, we have had the fuse plugs with the primer already, screwed in and in position before pouring in the powder (J) which was the last thing done before locating them and unscrewing the shields just previous to locating. This is probably the best torpedo for destruction of ships every made.—G.J.Rains

19. See manuscript page 65.

20. See October 5, 1863, Appendix 1.

21. Rains' 1873 patent for the "Improvement in Safety Valves" may be viewed at: http://www.google.com/patents?id=D4NKAAAAEBAJ&printsec=abstract&zoom=4&source=gbs_overview_r&cad=0#v=onepage&q=&f=false

22. Emile Lamm, "Fireless Locomotive," *Scientific American*, vol. XXVII August 24, 1872, p. 118.

23. This account differs somewhat from that Tomb wrote later for his family. R. Thomas Campbell (editor), *Engineer in Gray: Memoirs of Chief Engineer James H. Tomb, CSN* (Jefferson, NC: McFarland & Company, 2005), pp. 66–75.

24. "The Fastest Steamer in the World," *Scientific American*, Vol. XXX, No. 30, May 30, 1874, p. 343.

II. Peter S. Michie

Biographical Introduction

1. James H. Wilson, "Peter Smith Michie," *1901 Annual Reunion of the Association of Graduates, United States Military Academy*, pp. 149–150.

2. Wilson, pp. 151–152.

3. Wilson, pp. 153–154; *Congressional Series Set*, Senate Report No. 307 (Washington, Government Printing Office, 1901).

4. Wilson, pp. 157–158; George W. Cullum, *Biographical Register of the Officers and Graduates of the U.S. Military Academy, from 1802 to 1867* (New York: James Miller, 1879), p. 587.

5. Cullum, pp. 587–588; Wilson, p. 159; *Congressional Serial Set*, Senate Report 307 (Washington: Government Printing Office, 1901).

6. Wilson, pp. 160–162; *ORA*, Series 1, vol. 28, part 1, pp. 17, 335–340.
7. Cullum, p. 588; Wilson, p. 159.
8. Wilson, pp. 162–163; Cullum, p. 588.
9. Michie Papers, Letter of February 23, 1864 [1865], United States Military Academy; *ORA*, Series 1, vol. 42, part 1, pp. 657–673.
10. Cullum, p. 588; Wilson, p. 164; *ORA*, Series 1, vol. 42, part 3, pp 98, 167. 351.
11. Francis B. Heitman, *Historical Register of the United States Army* (Washington, D.C., The National Tribune, 1890), p. 464.
12. Cullum, p. 588; Wilson, pp. 166–167; ORA, Series 1, vol. 42, part 1, pp. 1165–1167.
13. Wilson, p. 168.
14. John Y. Simon (editor), The Papers of Ulysses S. Grant (Carbondale: Southern Illinois University Press, 1985), vol. 14, pp. 109–110, 154.
15. Wilson, p. 170; Cullum, p. 588; *ORA*, Series 1, vol. 46, part 3, pp. 1143–1144. In the *Atlas to Accompany the Official Records of the Union and Confederate Armies* (Washington, Government Printing Office, 1891–1895) are eleven of Michie's maps: Bermuda Hundred, Cobb's Hill, Cold Harbor, Deep Bottom, Dutch Gap Canal, Harrison's Landing, James River to New Market Road, Petersburg, and Richmond, Virginia, as well as Morris Island, S.C., and New Bern, N.C. In addition, several of the photographs Michie submitted of various batteries and structures are also included in the *Atlas*.
16. John Y. Simon (editor), *The Papers of Ulysses S. Grant* (Carbondale: Southern Illinois University Press, 1988), vol. 16, p. 20; Wilson, p. 171; James A. Padgett, "Reconstruction Letters from North Carolina" Part IX Letters to Benjamin Franklin Butler, *North Carolina Historical Review*, vol. 19, p. 402–404; continued in vol. 20, pp. 58–61. Michie's quarterly financial reports from April 1866 to April 1867 may be found in vol. 20, pp. 63–69.
17. Michie Papers, letter December 20, 1878, United States Military Academy.
18. *Index of General Orders, Adjutant General's Office, 1871* (Washington: Government Printing Office, 1872), p. 10.
19. Michie Papers, Letter December 20, 1878; Princeton University database:.https://webdb.princeton.edu/dbtoolbox/query.asp?qname=honorary&LNAME=michie&FNAME=peter&YEAR=&submit=++Submit++; personal communication from Jeanne M. West, Associate Dean for External Relations, Thayer School of Engineering at Dartmouth College, July 22, 2009.
20. Wilson, p. 177.
21. Wilson, p. 177.
22. Wilson, pp. 173–175.
23. Wilson, p. 178. Sadly, his widow Maria had no property, "real or personal, except some household effects of small value, and her only income is her pension." She relied "in great part" on her 28-year-old daughter for support, who lent "such assistance as she can offer from the very limited means at her disposal." The senate voted to increase the widow Michie pension from $30.00 to $40.00 a month (*Congressional Serial Set*, Senate Report No. 307) ; *New York Times*, February 22, 1901, p. 4.

Editorial Notes

1. W.R. King, *Torpedoes: Their Invention and Use, from the First Application to the Art of War to the Present Time*, Washington, Government Printing Office, 1866, pp. 7–18. A footnote on page 7 states, "the torpedoes only will be here given."
2. Comstock's published report, including the more rudimentary diagram of the fuse, may be found in *ORA*, Series 1, Vol. 46, part 2, pp. 215–217.

In June, 1865, Admiral John A. Dahlgren prepared a detailed report of torpedo operations in and around Charleston, S.C., based on his experience and interviews with various Confederates. His report, along with many sketches, can be found in *ORN*, Series 1, Vol. 16, pp. 380–406.

Notes Explaining Rebel Torpedoes and Ordnance as Shown in Plates Nos. 1 to 21 Inclusive

1. Elliot Lacy was on the staff of the Confederate Ordnance Department as an Assistant Inspector.

I assume John F. Alexander worked in the Torpedo Bureau, since detailed formulas for explosives are included on some of the plates and in the text. He may have been John Fielding Alexander who served in various Confederate artillery units, was hospitalized in late summer of 1862, and resigned in February 1863. No further record of military service exists; after the war he worked as a civil and mining engineer. (VMI Archives Online Rosters Database).

2. George Washington Rains, brother of Gabriel James Rains, was commander of the Confederate Powderworks at Augusta, Ga.
3. Plate VII lists 45 parts chlorate potassa.

Editor's Appendix 1

1. Gabriel Rains, in his *Torpedo Book*, claims that 58 vessels were sunken (see Rains manuscript p. 41).
2. *ORN*, Series 1, vol. 12, pp. 502–503. The report says the torpedoes contained 30 pounds of gunpowder; the diagram says 70 pounds of gunpowder. There is no evidence that the torpedoes in the Savannah River were galvanic torpedoes controlled from Fort Pulaski. For a detailed discussion of the preparations for the siege of Fort Pulaski, see Herbert M. Schiller, *Sumter is Avenged!: the Siege and Reduction of Fort Pulaski* (Shippensburg: White Mane Publishing Company, Inc., 1995).
3. *ORN*, Series 1, vol. 12, pp. 502–503.
4. *ORN*, Series 2, vol. 1, p. 49.
5. *ORN*, Series 1, vol. 23, p. 546–547; Perry, *Infernal Machines*, p. 32.
6. Isaac N. Brown, "Confederate Torpedoes in the Yazoo," *Battles and Leaders of the Civil War*, vol. 3 (New York: The Century Co., 1884), p. 580; Perry, *Infernal Machines*, p. 31.

7. *ORN*, Series 1, vol. 26, pp. 545, 548.
8. Perry, *Infernal Machines*, pp. 31–32; *ORN*, Series 1, vol. 23, pp. 545, 547–548.
9. Perry, *Infernal Machines*, p. 33, *ORN*, Series 1, vol. 23, pp. 547, 549–550; Series 2, p. 49.
10. Perry, *Infernal Machines*, p. 34; ORN Series 1, vol. 23, pp. 545, 547, 551.
11. Perry, *Infernal Machines*, p. 206.; http://www.nps.gov/vick/u-s-s-cairo-gunboat.htm
12. *ORN*, Series 1, vol. 23, p. 545.
13. http://www.history.navy.mil/photos/sh-usn/usnsh-k/kinsman.htm
14. *ORN*, Series 2, vol. 1, p. 122; Paul H. Silverstone, *Warships of the Civil War Navies* (Annapolis: Naval Institute Press, 2001), p. 72.
15. *ORN*, Series 1, vol. 19, pp. 521–522.
16. *ORN*, Series 1, vol. 19, pp. 518–519, 521.
17. ORN, Series 2, vol. 1, p. 149.
18. Perry, *Infernal Machines*, pp. 38–39.
19. *ORN*, Series 1, vol. 13, p. 630. The *Montauk* received fourteen hits on January 27 and 48 hits on February 1. (Silverstone, *Civil War Navies*, p. 6.)
20. *ORN*, Series 1, vol. 13, pp. 698, 700–704; Schafer, *Confederate Underwater Warfare*, p. 79.
21. *ORN*, Series 2, vol. 1, pp. 192–193.
22. *ORN*, Series 1, vol. 19, pp. 665–669.
23. *ORN*, Series 1, vol. 19, pp. 672–673. Eventually, only the *Hartford* and *Albatross* would pass above Port Hudson; the remainder of the vessels had to fall back.
24. E. Milby Burton, *The Siege of Charleston 1861–1865* (Columbia: University of South Carolina Press, 1970), pp. 22, 25; *The Charleston Mercury*, April 7, 1863, page 2, column 1; Silverstone, *Civil War Navies*, p. 182.
25. Perry, *Infernal Machines*, p. 52; *OR*, Series 1, vol. XVIII, p. 1082; *ORN*, Series 1, vol. 16, pp. 386, 402, 412; W. Craig Gains, *Encyclopedia of Civil War Shipwrecks* (Baton Rogue: Louisiana State University Press, 2008), p. 151; John T. Bucknill, *Submarine Mines and Torpedoes: As Applied to Harbour Defence* (New York: John Wiley & Sons, 1889), p. 2.
26. *ORN*, Series 1, vol. 16, p. 412. Confederate Captain M. Martin Gray, on General Gabriel Rains' staff and involved in the torpedo manufacturing facility in Charleston, reported that the *Ettiwan* was hit by a floating Confederate torpedo the morning after the *Marion* sank.
27. *ORN*, Series 1, 16, pp. 386, 412;.John T. Scharf, *History of the Confederate States Navy from Its Organization to the Surrender of Its Last Vessel*, 2nd edition (Albany: Joseph McDonough, 1894), p. 757.
28. The *Ettiwan* was also known as the *Etwan*, *Etowah*, *Etowan*, and *Hetiwan*. (Arthur Wyllie, *The Confederate States Navy* [Whitefish: Kessinger Pub. Co, 2007]), p. 49.
29. Gains, *Civil War Shipwrecks*, p. 145; *ORN*, Series 1, vol. 13, p. 823; vol. 16, pp. 388, 412; Scharf, *Confederate Navy*, p. 757; Bucknill, *Submarine Mines and Torpedoes*, p. 2; Wyllie, *Confederate States Navy*, p. 49; Silverstone, *Civil War Navies*, p. 181. There is no mention of the damage to the *Ettiwan* in *The Charleston Mercury* April-August 1863 issues or *The Charleston Courier* April 1863 issues.
30. Burton, *Siege of Charleston*, p. 231.
31. Gains, *Civil War Shipwrecks*, p. 145.
32. *ORN*, Series 2, vol. 1, p. 238.
33. Perry, *Infernal Machines*, p. 51.
34. *The Charleston Mercury*, April 8, 1863, page 1, column 2; *OR*, Series 1, vol. XIV, pp. 241, 246–250, 1017; Perry, *Infernal Machines*, p. 51.
35. Perry pp. 51–52; Schafer, *Confederate Underwater Warfare*, pp. 89–90.
36. *OR*, Series 1, vol. XIV, p. 1017.
37. *ORN*, Series 2, vol. 1, p. 42.
38. *ORN*, Series 1, vol. 25, pp. 146, 280, 281.
39. *ORN*, Series 1, vol. 25, p. 282, 283, 286; *ORN*, Series 2, vol. 1, p. 208. The Mississippi river boats were the *Magenta*, *Magnolia*, *Prince of Wales*, *Pargoud*, and *Peytona*.
40. *ORN*, Series 1, vol. 25, p. 282, 286.
41. *ORN*, Series 2, vol. 1, p. 42.
42. *ORN*, Series 1, vol. 25, p. 282; Brown, "Confederate Torpedoes on the Yazoo River," p. 580; Perry, *Infernal Machines*, p. 42.
43. *ORN*, Series 1, vol. 25, pp. 287–288. The vessels burned were the *Mary Keene*, *John Walsh*, *Lockland*, *Scotland*, *Golden Age*, *Arcadia*, *Kennett*, *Gay*, *Natchez*, and *Parallel* on the Yazoo River and the *Dewdrop*, *Emma Bett*, *Sharp*, and *Meares* on the Sunflower River.
44. *ORN*, Series 1, vol. 25, pp. 282, 283, 286–288.
45. W. Craig Gaines, *Encyclopedia of Civil War Shipwrecks* (Baton Rouge: Louisiana State University Press, 2008), p. 82.
46. *ORN*, Series 2, vol. 1, p. 62.
47. *ORN*, Series 1, vol. 9, pp. 145–148.
48. R.O. Crowley, "The Confederate Torpedo Service," *The Century Magazine*, Vol. 56 (May to October, 1898), p. 292.
49. *ORN*, Series 2, vol. 1, 172.
50. *ORN*, Series 1, vol. 14, p. 445.
51. *ORN*, Series 1, vol. 14, pp. 445–446, 448.
52. Schafer, *Confederate Underwater Warfare*, pp. 95–96.
53. *ORN*, Series 2, vol. 1, p. 159.
54. *ORN*, Series 1, vol. 15, pp. 12–13, 16; W.T. Glassell, "Torpedo Service in Charleston Harbor," Southern Historical Society Papers, vol. 4, July-December 1877, p. 229.
55. *ORN*, Series 1, vol. 15, p. 20–21. A more detailed account of James Tomb's experience can be found in Campbell, *Engineer in Gray*, pp. 65–75.
56. *ORN*, Series 1, vol. 15, p. 14.
57. *ORN*, Series 1, vol. 15, p. 19.
58. Schafer, *Confederate Underwater Warfare*, p. 100.
59. *ORN*, Series 2, vol. 1, p. 104.
60. *ORN*, Series 1, vol. 15, p. 329. Silverstone, *Civil War Navies*, p. 26, claims 5 men were lost.
61. *ORN*, Series 1, vol. 15, p. 330.
62. *ORN*, Series 1, vol. 15, p. 332.
63. *ORN*, Series 1, vol. 15, p. 331, 333.
64. *ORN*, Series 1, vol. 15, p. 335.
65. There are many accounts of the *Hunley* and her recovery. Representative is Brian Hicks and Schuler Kropf, *Raising the Hunley: The Remarkable History and Recovery of the Lost Confederate Submarine* (New York: Ballantine Books, 2002).
66. Gaines, *Civil War Shipwrecks*, p. 149.

67. Keith V. Holland, *et. al., The Maple Leaf: An Extraordinary American Civil War Shipwreck* (Jacksonville: St. Johns Archaeological Expeditions, Inc., 1993), pp. 92–107.

68. Perry, *Infernal Machines*, p. 115.

69. *New York Times*, April 13, 1864, page 1, column 1; Gaines, *Civil War Shipwrecks*, p. 42.

70. *ORA*, Series 1, vol. 35, pt. 1, p. 370, 380; *New York Times*, April 13, 1864, page 1, column 1.

71. *ORA*, Series I, vol. 35, pt. 1, p. 115.

72. *ORA*, Series 1, vol. 35, pt. 1, pp. 370, 381; *ORN*, Series 1, vol. 16, pp. 424–425; Gaines, *Civil War Shipwrecks*, p. 42; Schafer, *Confederate Underwater Warfare*, p. 130.

73. Holland, *The Maple Leaf*, pp. 92–107, 127–138, 145–147.

74. *ORN*, Series 2, vol. 2, p. 145.

75. *ORN*, Series 1, vol. 9, p. 954.

76. *ORN*, Series 1, vol. 9, pp. 593, 954, 599–600 (with details of damage), 603; Davidson, *Southern Historical Society Papers*, vol. 24, p. 286; Scharf, *Confederate Navy*, p. 762; Schafer, *Confederate Underwater Warfare*, p. 138. Scharf claims Davidson's steam torpedo-boat was the *Torpedo*.

77. Davidson, *Southern Historical Society Papers*, vol. 24, p. 286.

78. *ORN*, Series 1, vol. 9, pp. 602–603.

79. *ORN*, Series 2, vol. 1, p. 77.

80. Thomas O. Selfridge, "The Navy in the Red River," *Battles and Leaders of the Civil War*, vol. 4 (New York, The Century Co., 1888), pp. 364, 366.

81. *ORN*, Series 1, vol. 26, p. 78, 68–76. Lieutenant Commander Phelps reported that his launches, prior to the *Eastport* striking the torpedo, had "burst" three Confederate torpedoes in the vicinity without damage to the small boats.

82. O*RN*, Series 1, vol. 26, pp. 78, 68–76.

83. *ORN*, Series 1, vol. 26, pp. 79, 68–76.

84. Gains, *Civil War Shipwrecks*, p. 64.

85. *ORA*, Series 1, vol. 35, pt. 1, pp. 387–388; Perry, *Infernal Machines*, p. 116; Gaines, *Civil War Shipwrecks*, p. 40; *New York Times*, April 27, 1864, page 1, column 5.

86. Gaines, *Civil War Shipwrecks*, p. 40.

87. *ORN*, Series 2, vol. 1, p. 63.

88. *ORN*, Series 1, vol. 9, pp. 724–725, 729.

89. *ORN*, Series 1, vol. 10, pp. 9, 12, 15.

90. Hunter Davidson, *Confederate Veteran*, vol. XVI, p. 457; Richard H. Stothard, *Notes on Defence by Submarine Mines* (Brompton: James Gale, 1873), p. 3.

91. *ORN*, Series 1, vol. 10, pp 9, 14, 16, 26.

92. Shelby Foote, *The Civil War—A Narrative*, vol. 3 (New York: Random House, 1974), pp. 253–254.

93. Barnes, *Submarine Warfare*, p. 99.

94. *ORN*, Series 1, vol. 10, pp. 10, 14, 15.

95. *ORN*, Series 1, vol. 10, p. 14.

96. Silverstone, *Civil War Navies*, p. 70.

97. *ORN*, Series 1, vol. 9, p. 27; vol. 10, p. 3.

98. *New York Times*, May 18, 1864, page 5, column 1; *ORA*, Series 1, vol. 35, p. 392; Gaines, *Civil War Shipwrecks*, p. 41.

99. Richard A. Martin, "The Great River War on the St. Johns," http://www.mapleleafshipwreck.com/Book/Chapter2/chapter2.htm; *ORA*, Series 1, vol. 35, p. 226; Schafer, *Confederate Underwater Warfare*, p. 133.

100. *ORN*, Series 2, vol. 1, p. 221; Silverstone, *Civil War Navies*, p. 7.

101. *ORN*, Series 1, vol. 21, pp. 405, 417.

102. *ORN*, Series 1, vol. 21, p. 405.

103. *ORN*, Series 1, vol. 21, p. 569. For a detailed discussion on period coastal defense, see Viktor von Scheliha, *A Treatise on Coast-Defense* (London: E. & F. N. Spon, 1868).

104. *ORN*, Series 1, vol. 21, pp. 405, 569.

105. *ORN*, Series 1, vol. 21, p. 569.

106. *ORN*, Series 1, vol. 21, p. 570; *ORA*, Series 1, vol. 35, pp. 739, 759.

107. Silverstone, *Civil War Navies*, p. 7; Schafer, *Confederate Underwater Warfare*, p. 148; Casandra Andews, Mobile *Press-Register*, August 5, 2001; http://www.history.navy.mil/photos/sh-usn/usnsh-t/tecumseh.htm.

108. *ORA*, Series 1, vol. 42, part. 1, pp. 954–955; Gabriel J. Rains Papers, undated, South Carolina Historical Society; John Wideman, "The Confederate Sabotage Operation Against City Point, Virginia August 1864," Personal communication, December 11, 2009; Morris Schaff, "The Explosion at City Point," *Civil War Papers Read Before the Commandery of the State of Massachusetts, Military Order of the Loyal Legion of the United States* (Boston, 1900), vol. 2, p. 478.

109. Schaff, p. 480.

110. Schaff, p. 483.

111. Horace Porter, *Campaigning with Grant* (New York, The Century Company, 1907), p. 273.

112. Frederick H. Dyer, *A Compendium of the War of the Rebellion*, Pt. 2 (Des Moines: The Dyer Publishing Co., 1908), p. 950; Schaff, p. 278; Porter, p. 274. The *New York Times*, August 13, 1864, page 2, column 6, lists the names of 47 killed and 66 wounded.

113. *ORA*, Series 1, vol. 42, part 1, pp. 954–955.

In an undated three page document, Rains records the transcript of an article in the *New York Herald* (also undated), stating the report of Maxwell had been reviewed by J. Kellogg, A.A.G., and that when the true circumstances of the City Point explosion were learned by Federal authorities, Rains reported that "the design was to arrest and try me but my death being falsely published by the papers arrested the proceedings." (Gabriel J. Rains Papers, undated, South Carolina Historical Society). For a time, Federal authorities also searched, unsuccessfully, for John Maxwell (Kochen and Wideman, *Torpedoes*, p. 85). Peter S. Michie was aware that a horological torpedo had been used by the time he made his report in October 1865 (see manuscript p. 68).

114. David D. Porter, *Incidents and Anecdotes of the Civil War* (New York, D. Appleton and Company, 1886), pp. 263–266; *New York Times*, November 20, 1864, page 1, column 2.

115. Scharf, *Confederate Navy*, p. 762.

116. Barnes, *Submarine Warfare*, p. 76.

117. *ORN*, Series 2, vol. 1, p. 155.

118. *ORN*, Series 1, vol. 21, pp. 752–754.

119. *ORN*, Series 2, vol. 1, pp. 43, 167.

Scharf, *Confederate Navy*, pp. 766, 768, lists "Picket Boat No. 5" as a separate boat sunk on the Roanoke

River on December 10, 1864. The *New York Times* (December 17, 1864, page 1, column 4) discusses the sinking of the *Otsego* and adds "The picket-boat *Basly* [sic] No. 5 met with the same fate." There are no references to *Picket Boat No. 5* sinking on the Roanoke River along with the *Otsego* and *Bazely* in the *OR* or *ORN*; I believe the *Bazely* and *Picket Boat No. 5* were the same vessel. Gaines (*Civil War Shipwrecks*, p. 114) claims that the *Bazely* was also known as *Picket Boat No. 2*, but this is at variance with *ORN*, Series 2, vol. 1, p. 178.

120. William P. Derby, *Bearing Arms in the Twenty-seventh Massachusetts Regiment of Volunteer Infantry During the Civil War*. (Boston: Wright & Potter, 1883), pp. 446, 448–452.

111. Derby, pp. 447–449, 451; Perry, *Infernal Machines*, p. 150.

122. *ORN*, Series 1, vol. 11, p. 172; Perry, *Infernal Machines*, pp. 148–149.

123. *ORN*, Series 1, vol. 11, p. 162; *New York Times*, December 14, 1864, page 1, column 2.

124. *ORN*, Series 1, vol. 11, p. 164. Rains claims demijohn torpedoes were used (*Torpedo Book*, manuscript p. 25); Schafer, *Confederate Underwater Warfare*, p. 164, claims they were Fretwell-Singer torpedoes.

125. *ORN*, Series 1, vol. 11, p. 165.

126. *ORN*, Series 1, vol. 11, p. 163.

127. Barnes, *Submarine Warfare*, p. 108.

128. *ORN*, Series 1, vol. 11, p. 167.

129. *ORN*, Series 1, vol. 11, p. 167; *ORA*, Series, 1, vol. 42, part 3, p. 1050.

130. *ORA*, Series 1, vol. 42, part 3, p. 1050; *ORN*, Series 1, vol. 11, pp. 172–172.

131. *ORN*, Series 1, vol. 11, pp. 172–173.

132. *ORN*, Series 1, vol. 11, p. 171; *ORN*, Series 2, vol. 1, p. 168. Gabriel Rains mistakenly thought six vessels had been destroyed in the Roanoke River (*ORA*, Series, 1, vol. 47, part 3, p. 729, and Rains, *Torpedo Book*, manuscript p. 25.)

133. *ORN*, Series 2, vol. 1, pp. 170–171.

134. *ORN*, Series 1, vol. 16, p. 172.

135. *ORN*, Series 1, vol. 16, p. 174.

136. *ORN*, Series 1, vol. 16, pp. 173–178.

137. *ORA*, Series 1, vol. 47, pt. 1, p. 1068.

138. *ORA*, Series 1, vol. 47, pt. 1, p. 1036.

139. *ORN*, Series 1, vol. 16, p. 174; James H. Tomb, "The Last Obstructions in Charleston Harbor, 1863," *Confederate Veteran*, vol. XXXII, 1924, p. 98.

140. Gaines, *Civil War Shipwrecks*, p. 153.

141. John M. Coski, *Capital Navy* (New York, Savas Beatie LLC, 2005), p. 212; Wyllie, *Confederate States Navy*, p. 147; Scharf, *Confederate Navy*, p. 767; Silverstone, *Civil War Navies*, p. 243.

142. Coski, p. 287; Hunter Davidson, *Confederate Veteran*, vol. XVI, p. 459; Davidson, *Southern Historical Society Papers*, vol. 24, p. 287; Scharf, *Confederate Navy*, p. 767; *New York Times*, February 19, 1865, page 4, column 5; Schafer, *Confederate Underwater Warfare*, p. 39. The *Times* misidentified the *Schultz* as the 301-ton screw steamer *William Alison*, claims she "went down almost immediately," and that there were no survivors.

143. *ORN*, Series 1, vol. 12, p. 186.

144. *ORA*, Series 1, vol. 46, pt. 2, p. 646.

145. Chris E. Fonvielle, Jr., *The Wilmington Campaign* (Campbell: Savas Publishing Co., 1997), pp. 389, 399–400.

146. Porter, *Incidents and Anecdotes*, p. 278. In his report dated February 22, 1865, Admiral Porter reported "the rebels sent down 200 floating torpedoes, but I had a strong force of picket boats out, and the torpedoes were sunk with musketry.... Some of the vessels picked up torpedoes with their torpedo nets." *New York Times*, February 26, 1865, page 1, column 1.

147. *ORN*, Series 1, vol. 12, pp. 37, 44; Porter, *Incidents and Anecdotes*, p. 278.

148. *ORN*, Series 2, vol. 1, p. 167.

149. Fonvielle, *Wilmington Campaign*, pp. 387–424.

150. *ORN*, Series 1, vol. 12, p. 45.

151. *ORN*, Series 2, vol. 1, p. 99.

152. *ORN*, Series 1, vol. 16, p. 283.

153. *ORN*, Series 1, vol. 16, pp. 282–283.

154. Gaines, *Civil War Shipwrecks*, p. 147.

155. Charles D. Gibson and E. Kay Gibson, *The Army's Navy Series: Dictionary of Transports and Combatant Vessels Steam and Sail Employed by the Union Army, 1861–1865* (Camden: Ensign Press, 1995), p. 315; John McElroy, *Andersonville* (Washington: The National Tribune, 1913), p. 261–263; *New York Times*, March 16, 1865, page 11, column 1; Scharf, *Confederate Navy*, p. 767; *ORA*, Series, 1, vol. 47, part 3, p. 729.

156. *ORN*, Series 2, vol. 1, p. 115.

157. *ORN*, Series 1, vol. 16, p. 297.

158. *ORN*, Series 1, vol. 16, pp. 296–297, 408–409.

159. Bern Anderson, *By Sea and By River: The Naval History of the Civil War* (New York: Alfred A. Knopf, 1962), pp. 248–249. An excellent account of the final operations of the Mobile campaign can by found in Chester G. Hearn, *Mobile Bay and the Mobile Campaign* (Jefferson, NC: McFarland, 1993), pp. 146–201.

160. *ORN*, Series 2, vol. 1, p. 33.

161. *ORN*, Series 1, vol. 22, p. 133.

162. Gaines, *Civil War Shipwrecks*, p. 1.

163. Silverstone, *Civil War Navies*, p. 146.

164. *ORN*, Series 1, vol. 16, pp. 295–296.

165. *ORN*, Series 2, vol. 1, p. 144.

166. *ORN*, Series 1, vol. 22, p. 71; James R. Soley, "Closing Operations in the Gulf and Western Rivers," *Battles and Leaders of the Civil War*, vol. 4 (New York: The Century Co., 1884), p. 412.

167. *ORN*, Series 2, vol. 2, p. 167.

168. *ORN*, Series 1, vol. 22, pp. 72–73.

169. Gaines, *Civil War Shipwrecks*, p. 5.

170. *ORN*, Series 2, vol. 1, p. 194; Gaines, *Civil War Shipwrecks*, p. 6. In *ORN*, Series 1, vol. 16, p. 432, in an October 16, 1865, letter Admiral John Dahlgren lists *Tinclad No. 48*, in addition to the *Rodolph*, being sunk in Mobile Bay; he did not serve in the West Gulf Blockading Squadron and seems to have confused the two names of the vessel.

171. *ORN*, Series 1, vol. 22, pp. 72–73; Gaines, *Civil War Shipwrecks*, p. 6.

172. *ORN*, Series 2, vol. 1, p. 106.

173. *ORN*, Series 1, vol. 22, p. 131; *ORN*, Series 2, vol. 1, p. 106.

174. Gaines, *Civil War Shipwrecks*, p. 3.

175. *ORN*, Series 2, vol. 1, p. 203.

176. *ORN*, Series 1, vol. 22, p. 130.
177. Gaines, *Civil War Shipwrecks*, p. 6.
178. *ORN*, Series 1, vol. 22, pp. 130–131.
179. *New York Times*, May 7, 1865, p. 1, column 2; Scharf, *Confederate Navy*, p. 767. Scharf dates the sinking on May 12, 1865. The story in the May 7, 1865, *New York Times* cites the April 27, 1865, *New Orleans Delta*.
180. "Absence of proof is not proof of absence." Source material may exist, but concerning the *John Farron*, *Rose*, and *St. Mary's*, I was unable to locate it in official publications, period newspapers, or secondary sources.
181. Although the *John Farron* is **listed** as seriously damaged in King, *Torpedoes*, p. 91, and repeated in lists by Scharf, *Confederate Navy*, p. 768, and in identical lists in John T. Bucknill, *Submarine Mines*, p. 2 and *Journal of the Military Service Institution of the United States*, vol. 1, 1880, p. 205. Schafer, *Confederate Underwater Warfare*, p. 113, states the damage was the result of a galvanic torpedo, but cites no reference. There is no listing of damage by torpedoes in *ORA*, *ORN*, or in Charles D. Gibson and E. Kay Gibson, *The Army's Navy Series: Dictionary of Transports and Combatant Vessels Steam and Sail Employed by the Union Army, 1861–1865* (Camden: Ensign Press, 1995), p. 178.
182. *American Lloyds' Registry of American and Foreign Shipping* (New York: Ferris & Pratt, 1862), pp. 574–575.
183. *ORN*, Series 2, vol. 1, p. 140.
184. *ORN*, Series 1, vol. 15, p. 358.
185. *ORN*, Series 1, vol. 15, p. 359.
186. *ORN*, Series 1, vol. 15, p. 359.
187. *ORN*, Series 1, vol. 15, pp. 356–357.
188. *ORN*, Series 1, vol. 15, p. 359; Schafer, *Confederate Underwater Warfare*, pp. 100–101.
189. Scharf, *Confederate Navy*, p 767. There is no listing of *Rose* sinking by torpedoes in *ORA*, *ORN*, or Gibson, *The Army's Navy Series*, p. 178.
190. *ORN*, Series 2, vol. 1, p. 195.
191. Scharf, *Confederate Navy*, p. 767. Scharf claims the *St. Mary's* sank a "a few days after" the *Rose*. There is no listing of *St. Mary's* sinking by torpedoes in *OR*, *ORN*, or Gibson, *The Army's Navy Series*, p. 178.
192. *ORN*, Series 2, vol. 1, p. 138.
193. *ORN*, Series 1, vol. 16, p. 296.

Bibliography

Manuscripts

National Archives

Microfilm Publication 2048. War Department Letters and Telegrams Sent Relating to the United States Military Academy, 1867–1904. Roll 1, Letters and Telegrams Sent. Vol. 3: July 1, 1874–October 13, 1875.

The Museum of the Confederacy

Brockenbrough, Eleanor S., letter November 1970.

Rains, Gabriel J., *Torpedo Book.*

South Carolina Historical Society

Gabriel J. Rains Papers

United States Military Academy

Gabriel J. Rains Papers

Peter S. Michie, *Notes Explaining Rebel Torpedoes and Ordnance as Shown in Plates Nos. 1 to 21 Inclusive.*

Peter Smith Michie Papers

Chris E. Fonvielle, Jr., personal communication, July 17, 2009.

Jeanne West, personal communication, July 22, 2009.

John Wideman, "The Confederate Sabotage Operation Against City Point, Virginia August 1864," Personal communication, December 11, 2009

Books, Articles, Theses, Newspapers, Databases

Anderson, Bern. *By Sea and By River: The Naval History of the Civil War.* New York: Alfred A. Knopf, 1962.

American Lloyds' Registry of American and Foreign Shipping. New York: Ferris & Pratt, 1862.

Andrews, Casandra. "Tecumseh's Final Resting Place" Mobile *Press-Register*, August 5, 2001.

Atlas to Accompany the Official Records of the Union and Confederate Armies. Washington, DC: Government Printing Office, 1891–1895.

Barnes, John S. *Submarine Warfare, Offensive and Defensive, Including a Discussion of the Offensive Torpedo System, Its Effects Upon Iron-Clad Ship Systems, and Influence Upon Future Naval Wars.* New York: D. Van Nostrand, 1869.

Barrett, John G. *The Civil War in North Carolina.* Chapel Hill: University of North Carolina Press, 1963.

Bell, Jack. *Civil War Heavy Explosive Ordnance.* Denton: University of North Texas Press, 2003.

Brown, Isaac N. "Confederate Torpedoes in the Yazoo." *Battles and Leaders of the Civil War*, volume 3. New York: The Century Company, 1884), p. 580.

Bucknill, John T. *Submarine Mines and Torpedoes: As Applied to Harbour Defence.* New York: John Wiley & Sons, 1889.

Burton, E. Milby. *The Siege of Charleston 1861–1865.* Columbia: University of South Carolina Press, 1970.

"U.S.S. *Cairo*," http://www.nps.gov/vick/u-s-s-cairo-gunboat.htm

Campbell, R. Thomas (editor). *Engineer in Gray: Memoirs of Chief Engineer James H. Tomb, CSN.* Jefferson, NC: McFarland, 2005.

Carraway, Gertrude. "North Caroliniana." Charlotte *Observer*, January 11, 1942.

The Charleston Mercury

Cullum, George W. *Biographical Register of the Officers and Graduates of the U.S. Military Academy, from 1802 to 1867.* New York: James Miller, 1879.

Congressional Series Set, Senate Report No. 307. Washington, DC: Government Printing Office, 1901.

Coski, John M. *Capital Navy.* New York: Savas Beatie LLC, 2005.

Crowley, R.O. "The Confederate Torpedo Service." *The Century Magazine*, volume 56 (May to October, 1898), pp. 290–300.

Davidson, Hunter. "A Chapter of War History Concerning Torpedoes." *Southern Historical Society Papers*, volume 24, 1895, pp. 284–281.

_____. "The Electrical Submarine Mine, 1861–1865." *Confederate Veteran*, volume XVI, 1908, pp. 456–459.

Delafield, Richard. *Report of the Art of War in Europe in 1854, 1855, and 1856.* Washington: George W. Bowman, 1860.

Derby, William P. *Bearing Arms in the Twenty-seventh Massachusetts Regiment of Volunteer Infantry During the Civil War.* Boston: Wright & Potter, 1883.

Dyer, Frederick H. *A Compendium of the War of the Rebellion*, Part 2. Des Moines: Dyer Publishing, 1908.

Fonvielle, Chris E., Jr. *The Wilmington Campaign.* Campbell: Savas Publishing, 1997.

Foote, Shelby. *The Civil War—A Narrative*, volume 3. New York: Random House, 1974.

Gains, W. Craig. *Encyclopedia of Civil War Shipwrecks.* Baton Rogue: Louisiana State University Press, 2008.

Gibson, Charles D., and E. Kay Gibson. *The Army's Navy Series: Dictionary of Transports and Combatant Vessels Steam and Sail Employed by the Union Army, 1861–1865.* Camden: Ensign Press, 1995.

Glassell, W.T. "Torpedo Service in Charleston Harbor." *Southern Historical Society Papers*, volume 4, July–December 1877, pp. 225–235.

Heitman, Francis B. *Historical Register of the United*

States Army. Washington, DC: *The National Tribune*, 1890.

Hearn, Chester G. *Mobile Bay and the Mobile Campaign*. Jefferson, NC: McFarland, 1993.

Hicks, Brian, and Schuler Kropf. *Raising the Hunley: The Remarkable History and Recovery of the Lost Confederate Submarine*. New York: Ballantine, 2002.

Holland, Keith V., et al. *The Maple Leaf: An Extraordinary American Civil War Shipwreck*. Jacksonville: St. Johns Archaeological Expeditions, 1993.

"The History of Landmines," http://members.iinet.net.au/~pictim@iinet.net.au/mines/history/history.html

Index of General Orders, Adjutant General's Office, 1871. Washington, DC: Government Printing Office, 1872.

Jones, J.B. *A Rebel War Clerk's Diary*. New York: Old Hickory Bookshop, 1935.

Journal of the Military Service Institution of the United States, volume 1, 1880, p. 205.

King, W.R. *Torpedoes: Their Invention and Use, from the First Application to the Art of War to the Present Time*. Washington, DC: Government Printing Office, 1866.

"U.S.S. *Kinsman*," http://www.history.navy.mil/photos/sh-usn/usnsh-k/kinsman.htm

Kochan, Michael P., and John C. Wideman. *Torpedoes: Another Look at the Infernal Machines in the Civil War*. Paoli: self published, n.d.

Lamb, William. "The Defense of Fort Fisher," in *Battles and Leaders of the Civil War*. New York: The Century Company, 1884, volume 4, pp. 642–654.

McElroy, John. *Andersonville*. Washington, DC: *The National Tribune*, 1913.

Nowlin, S.H. "Capture and Escape of S.H. Nowlin, Private Fifth Virginia Cavalry." *The Southern Bivouac*, volume 2, Oct. 1883, p. 70.

Official Records of the Union and Confederate Navies in the War of the Rebellion. Washington, DC: Government Printing Office, 1894–1914.

Padgett, James A. "'Reconstruction Letters from North Carolina.' Part IX Letters to Benjamin Franklin Butler," *North Carolina Historical Review*, volume 19, 1942, pp. 402–404, & volume 20, 1943, pp. 58–69.

Perry, Milton F. *Infernal Machines*. Baton Rogue: Louisiana State University Press, 1965.

Porter, David D. *Incidents and Anecdotes of the Civil War*. New York: D. Appleton, 1886.

Porter, Horace. *Campaigning with Grant*. New York: The Century Company, 1907.

Princeton University database: https://webdb.princeton.edu/dbtoolbox/query.asp?qname=honorary&LNAME=michie&FNAME=peter&YEAR=&submit=++Submit++

Pyne, Henry R. *History of the First New Jersey Cavalry*. Trenton: n.p., 1871.

Rains, Gabriel J. "Torpedoes." *Southern Historical Society Papers*, volume 3, 1887, pp. 255–260.

Rhea, Gordon C. *To the North Anna River*. Baton Rouge: Louisiana State University Press, 2000.

Sanford, George B. *Fighting Rebels and Redskins*. Norman: University of Oklahoma Press, 1964.

Scharf, John T. *History of the Confederate States Navy from Its Organization to the Surrender of Its Last Vessel*, 2nd edition. Albany: Joseph McDonough, 1894.

Schaff, Morris. "The Explosion at City Point." *Civil War Papers Read Before the Commandery of the State of Massachusetts, Military Order of the Loyal Legion of the United States*, volume 2, Boston, 1900, pp. 477–485.

Schafer, Louis C. *Confederate Underwater Warfare*. Jefferson, NC: McFarland, 1996.

von Scheliha, Viktor. *A Treatise on Coast-Defense*. London: E. & F. N. Spon, 1868.

Selfridge, Thomas O. "The Navy in the Red River." *Battles and Leaders of the Civil War*. New York: The Century Company, 1884, volume 4, pp. 362–366.

Sherman, William T. *The Memoirs of General William T. Sherman*. New York: The Library of America, 1990.

Silverstone, Paul H. *Warships of the Civil War Navies*. Annapolis: Naval Institute Press, 2001.

Simon, John Y., editor. *The Papers of Ulysses S. Grant*, volumes 14, 16, and 25, Carbondale: Southern Illinois University Press, 1985, 1988, and 2003.

Soley, James R., "Closing Operations in the Gulf and Western Rivers." *Battles and Leaders of the Civil War*. New York: The Century Company, 1884, volume 4, p. 412.

South, Stanley. *An Archaeological Education*. New York: Springer Scientific — Business Media, 2005.

Stothard, Richard H. *Notes on Defence by Submarine Mines*. Brompton: James Gale, 1873.

"U.S.S. *Tecumseh*," http://www.history.navy.mil/photos/sh-usn/usnsh-t/tecumseh.htm.

Tomb, James B. "The Last Obstructions in Charleston Harbor, 1863." *Confederate Veteran*, volume XXXII, 1924, pp. 98–99.

Virginia Military Institute Archives Online Rosters Database: http://www1.vmi.edu/archiverosters/Details.asp?ID=5&rform=search

The War of the Rebellion: A Compilation of the Official Records of the Union and Confederate Armies. Washington, DC: Government Printing Office, 1880–1901.

Wise, Stephen R. *Gate of Hell*. Columbia: University of South Carolina Press, 1994.

Waters, William Davis. "'Deception Is the Art of War': Gabriel J. Rains, Torpedo Specialists of the Confederacy." *The North Carolina Historical Review*, volume 66, number 1, 1989, pp. 29–60.

Waters, William Davis. "Gabriel J. Rains: Torpedo General of the Confederacy." Winston-Salem, Wake Forest University Masters Thesis, 1971.

Wilson, James H. "Peter Smith Michie." *1901 Annual Reunion of the Association of Graduates*. United States Military Academy, pp. 149–179.

Wolters, Timothy S. "Electric Torpedoes in the Confederacy: Reconciling Conflicting Histories." *Journal of Military History*, volume 72, pp. 755–783.

Wyllie, Arthur. *The Confederate States Navy*. Whitefish, MT: Kessinger, 2007.

Index

Page numbers in ***bold italics*** indicate pages with photographs.